T0141876

Studies in Systems, Decision and Control

Volume 71

Series editor

Janusz Kacprzyk, Polish Academy of Sciences, Warsaw, Poland
e-mail: kacprzyk@ibspan.waw.pl

About this Series

The series "Studies in Systems, Decision and Control" (SSDC) covers both new developments and advances, as well as the state of the art, in the various areas of broadly perceived systems, decision making and control- quickly, up to date and with a high quality. The intent is to cover the theory, applications, and perspectives on the state of the art and future developments relevant to systems, decision making, control, complex processes and related areas, as embedded in the fields of engineering, computer science, physics, economics, social and life sciences, as well as the paradigms and methodologies behind them. The series contains monographs, textbooks, lecture notes and edited volumes in systems, decision making and control spanning the areas of Cyber-Physical Systems, Autonomous Systems, Sensor Networks, Control Systems, Energy Systems, Automotive Systems, Biological Systems, Vehicular Networking and Connected Vehicles, Aerospace Systems, Automation, Manufacturing, Smart Grids, Nonlinear Systems, Power Systems, Robotics, Social Systems, Economic Systems and other. Of particular value to both the contributors and the readership are the short publication timeframe and the world-wide distribution and exposure which enable both a wide and rapid dissemination of research output.

More information about this series at http://www.springer.com/series/13304

António Pedro Costa · Luís Paulo Reis
Francislê Neri de Sousa · António Moreira
David Lamas

Editors

Computer Supported Qualitative Research

Springer

Editors
António Pedro Costa
Department of Education, Research Centre
  for Didactics and Technology in Teacher
  Education
University of Aveiro
Aveiro
Portugal

Luís Paulo Reis
School of Engineering, Department of
  Information Systems, Campus de Azurém
University of Minho
Guimarães
Portugal

Francislê Neri de Sousa
Department of Education, Research Centre
  for Didactics and Technology in Teacher
  Education
University of Aveiro
Aveiro
Portugal

António Moreira
School of Digital Technologies
Tallinn University
Tallinn
Estonia

David Lamas
School of Digital Technologies
Tallinn University
Tallinn
Estonia

ISSN 2198-4182   ISSN 2198-4190   (electronic)
Studies in Systems, Decision and Control
ISBN 978-3-319-82776-6   ISBN 978-3-319-43271-7   (eBook)
DOI 10.1007/978-3-319-43271-7

Preface

This book contains a selection of the papers accepted for presentation and discussion at first International Symposium on Qualitative Research (ISQR2016), held in Porto, Portugal, July 12–14, 2016. ISQR2016 is part of the Ibero-American Congress on Qualitative Research (CIAIQ). The ISQR/CIAIQ2016 conference organization had also the collaboration of several universities, research institutes and companies, including Lusófona University of Porto, University of Aveiro, University of Minho, University of Tallin, CIDTFF, LIACC, webQDA, Micro IO, Ludomedia, University of Tiradentes, CiberDid@tic Group of Extremadura University, Faculty of Psychology and Education Sciences of the University Porto, among several others.

ISQR2016 featured four main application fields (Education, Health, Social Sciences and Engineering and Technology) and seven main subjects: Rationale and paradigms of qualitative research (theoretical studies, critical reflection about epistemological dimensions, ontological and axiological); systematization of approaches with qualitative studies (literature review, integrating results, aggregation studies, meta-analysis, meta-analysis of qualitative meta-synthesis, meta-ethnography); qualitative and mixed methods research (emphasis in research processes that build on mixed methodologies but with priority to qualitative approaches); data analysis types (content analysis, discourse analysis, thematic analysis, narrative analysis, etc.); innovative processes of qualitative data analysis (design analysis, articulation and triangulation of different sources of data—images, audio, video); qualitative research in Web context (eResearch, virtual ethnography, interaction analysis, latent corpus on the internet, etc.); qualitative analysis with support of specific software (usability studies, user experience, the impact of software on the quality of research and analysis).

ISQR2016 featured one plenary talk on the state of the art of computer assisted qualitative data analysis, coordinated by Francislê Neri de Sousa, where Christina Silver, from the University of Surrey, Guildford, UK, presented "CAQDAS at a Crossroads: Controversies, Challenges and Choices". The conference also included two tutorial sessions, the first one promoted by Jaime Ribeiro, from the Polytechnic

Institute of Leiria, with the subject "Content Analysis Through Software (QDAS)" and the second one promoted by Brígida Mónica Faria from the Polytechnic Institute of Porto and Luis Paulo Reis from the University of Minho on the subject: "Data Mining Using RapidMiner for Mixed Methods".

CIAIQ2016 conference in total received 742 submissions from 29 countries. After a careful review process with at least three independent reviews for each paper, totally 50 high-quality papers from the papers submitted to the ISQR track were selected for publication in this Springer volume. The selected papers were written by 88 authors, from 17 countries, including Spain, Portugal, Brazil, Russian Federation, New Zealand, United Kingdom, United States, Israel, Ireland, Germany, France, Estonia, Czech Republic, Canada, Turkey, Rwanda, and Poland.

We would like to thank all CIAIQ and ISQR organizers and track chairs for their hard work on organizing and promoting the conference, inviting and managing the program committee, organizing the review process, and helping us to promote the conference. We would also like to thank the ISQR2016 program committee for their hard work on reviewing the papers and all ISQR2016 authors and participants for their decisive contribution to the success of the conference. A final word of thanks to all Springer staff for their help on producing this volume.

Aveiro, Portugal   António Pedro Costa
Guimarães, Portugal   Luís Paulo Reis
Aveiro, Portugal   Francislê Neri de Sousa
Tallinn, Estonia   António Moreira
Tallinn, Estonia   David Lamas

Contents

Research through Design: Qualitative Analysis to Evaluate the Usability

António Pedro Costa, Francislê Neri de Sousa, António Moreira
and Dayse Neri de Souza

Abstract The present need to identify and understand non-measurable/ non-quantifiable of the user experience with software has been the moto for may researchers in the area of Human Computer Interaction (HCI) to adopt qualitative methods. On the other hand, the use of qualitative analysis to support software has been growing. The integration of these types of tools in research is accompanied by an increase in the number of software packages available. Depending on the design and research questions, researchers can explore various solutions available in the market. Thus, it is urgent to ensure that these tools, apart from containing the necessary functionality for the purposes of research projects, are also usable. This study presents an assessment of the usability of the qualitative data analysis software webQDA® (version 2.0). To assess its usability, Content Analysis was used. The results indicate that the current version is "acceptable" in terms of usability, as users, in general, show a positive perception is perceived from users with completed PhDs as compared to those who are still doing their PhDs; no relevant differences can be found between the views obtained from different professional or research areas, although a more positive assessment may be drawn from Education and Teaching. Suggestions for future studies are put forward, even though we recognise that, in spite of most studies on usability defining quantitative metrics, the present study is offered as a contribution that aims to show that qualitative analysis has a great potential to deepen various dimensions of usability and functional and emotional interrelationships that are not by any means quantifiable.

A.P. Costa (✉) · F.N. de Sousa · A. Moreira · D.N. de Souza
Department of Education and Psychology, University of Aveiro, Aveiro, Portugal
e-mail: apcosta@ua.pt

F.N. de Sousa
e-mail: fns@ua.pt

A. Moreira
e-mail: moreira@ua.pt

D.N. de Souza
e-mail: Dayneri@ua.pt

© Springer International Publishing Switzerland 2017
A.P. Costa et al. (eds.), *Computer Supported Qualitative Research*,
Studies in Systems, Decision and Control 71,
DOI 10.1007/978-3-319-43271-7_1

1

Keywords Usability · Qualitative computing · Qualitative research · Qualitative data analysis · Research through design

1 Introduction

The evaluation of usability is much discussed, especially when we approach the graphic interfaces software. The study of usability is critical because certain software applications are "one-click" away from not being used at all, nor used in an appropriate manner. Usability comes as the most "rational" side of a product, allowing users to reach specific objectives in an efficient and satisfactory manner. Moreover, the User Experience is largely provided by the feedback on the usability of a system, reflecting the more "emotional" side of the use of a product. The experiment is related to the preferences, perceptions, emotions, beliefs, physical and psychological reactions of the user during the use of a product (ISO9241-210 2010). Thus, one comprehends "the pleasure or satisfaction" that many interfaces offer users, as evidence of efficiency in the integration of the concepts of Usability and User Experience in the development of HCI solutions (Scanlon et al. 2015). Despite being linked, in this paper we will focus the study on the dimension of Usability.

When dealing with authoring tools, in which you have to apply your knowledge to produce something, it becomes even more sensitive to gauge the usability of the software. Being webQDA® (Web Qualitative Data Analysis: www.webqda.net) an authoring tool for qualitative data analysis and a new version of it being developed (whose release is due April 2016) (Souza et al. 2016), it is of extreme relevance to keep this dimension in mind.

webQDA is a qualitative data analysis software that is meant to provide a collaborative, distributed environment (www.webqda.net) for qualitative researchers. Although there are some software packages that deal with non-numeric and unstructured data (text, image, video, etc.) in qualitative analysis, webQDA is a software directed to researchers in academic and business contexts who need to analyse qualitative data, individually or collaboratively, synchronously or asynchronously. webQDA follows the structural and theoretical design of other programs, with the difference that it provides online collaborative work in real time and a complementary service to support research (Souza et al. 2011).

In this paper, the main objective is to answer the following question: How to assess usability and functionality of the qualitative analysis software webQDA? After these initial considerations, it is important to understand the content of the following sections of this paper. Thus, in Sect. 2 we will present concepts associated with Qualitative Data Analysis Software. Sect. 3 will discuss Design through Research, while Sect. 4 will assess webQDA from the point of view of Usability, a methodological aspect where we present the methods and techniques for assessing the usability of software and the results of this study. And finally, we conclude with the study's findings.

2 Qualitative Data Analysis Software (QDAS)

The use of software for scientific research is currently very common. The spread of computational tools can be perceived through the popularization of software for quantitative and qualitative research. Nevertheless, nonspecific and mainly quantitative tools, like Word®, SPSS®, Excel®, etc., are the ones with major incidence or dissemination. This is also a reflection of the large number of books that can be found on quantitative research. However, the integration of specific software for qualitative research is a relatively marginal and more recent phenomenon.

In the context of postgraduate educational research in Brazil, some authors Teixeira et al. (2015) studied the use of computational resource in research. They concluded that among those who reported using software (59.9 %), there is a higher frequency of the use of quantitative analysis software (41.1 %), followed by qualitative analysis software (39.4 %), and finally the use of bibliographic reference software (15.5 %).

The so-called Computer-Assisted Qualitative Data Analysis Software or Computer Assisted Qualitative Data Analysis (CAQDAS) are systems that go back more than three decades (Richards 2002b). Today we can simply call them Qualitative Data Analysis Software (QDAS). However, even today many researchers are unaware of these specific and useful tools. Cisneros Puebla (2012) specifies at least three types of researchers in the field of qualitative analysis: (i) Researchers who are pre-computers, who prefer coloured pencils, paper and note cards; (ii) Researches who use non-specific software, such as word processors, spreadsheet calculations and general databases, and (iii) Researchers who use specific software to analyse qualitative research, such as NVivo, Atlas. it, webQDA, MaxQDA, etc.

We can summarize the story of specific qualitative research software with some chronological aspects:

(1) In 1966, MIT developed "The General Inquiry" software to help text analysis, but some authors (Tesch 1991; Cisneros Puebla 2003) refer that this was not exactly a qualitative analysis software.
(2) In 1984, the software Ethnograph comes to light, and still exists in its sixth version. (http://www.qualis.research.com/).
(3) In 1987, Richards and Richards developed the Non-Numerical Unstructured Data Indexing, Searching and Theorizing (NUD*IST), the software that evolved into the current NVivo index system.
(4) In 1991, the prototype of the conceptual network ATLAS-ti is launched, mainly related with Grounded Theory.
(5) Approximately in the transition of the 2000 decade it was possible to integrate video, audio and image in text analysis of qualitative research software. Nevertheless, HyperRESEARCH had been presented before as software that also allowed to encode and recover text, audio and video. Transcriber and Transana are other software systems that emerged to handle this type of data.

(6) In 2004, "NVivo summarizes some of the most outstanding hallmark previous software, such as ATLAS-ti—recovers resource coding in vivo, and ETHNOGRAPH—a visual presentation coding system" (Cisneros Puebla 2003).
(7) 2009 marks the beginning of the developments of qualitative software in cloud computer contexts. Examples of this are Dedoose and webQDA, that were developed almost simultaneously in USA and Portugal, respectively.
(8) From 2013 onwards we saw an effort from software companies to develop iOS versions, incorporating data from social networks, multimedia and other visual elements in the analysis process.

Naturally this story is not complete. We can include other details and software such as MaxQDA, AQUAD, QDA Miner, etc. For example, we can see a more exhaustive list in Wikipedia's entrance: "Computer-assisted qualitative data analysis software".

What is the implication of Qualitative Data Analysis Software on scientific research in general and specifically on qualitative research? Just as the invention of the piano allowed composers to begin writing new songs, the software for qualitative analysis also affected researchers in the way they dealt with their data. These technological tools do not replace the analytical competence of researchers, but can improve established processes and suggest new ways to reach the most important issue in research: to find answers to research questions. Some authors (Kaefer et al. 2015; Neri de Souza et al. 2011) recognize that QDAS allow making data visible in ways not possible with manual methods or non-specific software, allowing for new insights and reflections on a research or *corpus* of data.

Kaefer et al. (2015) wrote a paper with step-by-step QDAS software descriptions about the 230 journal articles they analysed about climate change and carbon emissions. They concluded that while qualitative data analysis software does not do the analysis for the researcher, "it can make the analytical process more flexible, transparent and ultimately more trustworthy" (Kaefer et al. 2015). There are obvious advantages in the integration of QDAS in standard analytical processes, as these tools open up new possibilities, such as agreed by Richards (2002b): (i) computers have enabled new qualitative techniques that were previously unavailable, (ii) computation has produced some influence on qualitative techniques. We therefore can summarize some advantage of QDAS: (i) faster and more efficient data management; (ii) increased possibility to handle large volumes of data; (iii) contextualization of complexity; (iv) technical and methodological rigor and systematization; (v) consistency; (vi) analytical transparency; (vii) increased possibility of collaborative teamwork, etc. However, many critical problems present challenges to the researchers in this area.

There are many challenges in the QDAS field. Some are technical or computational issues, whereas other are methodological or epistemological prerequisites, although Richards (2002a, b), recognized that methodological innovations are rarely discussed. For example, Corti and Gregory (2011) discuss the problem of exchangeability and portability of current software. They argue the need of data sharing, archiving and open data exchange standards among QDAS, to guarantee sustainability of data collections, coding and annotations on these data. Several researchers place expectations on the QDAS utilities on an unrealistic basis, while others think that the system has insufficient analytical flexibility. Richards (2002a) refers that many novice researchers develop a so-called "coding fetishism", that transforms coding processes into an end in itself. For this reason, some believe that QDAS can reduce critical reading and reflection. For many researchers, the high financial cost of the more popular QDAS is a problem, but in this paper we would like to focus on the challenge of the considerable time and effort required to learn them.

Choosing a QDAS is a first difficulty that, in several cases, is coincidental with the process of qualitative research learning. Kaefer et al. (2015) suggest to compare and test software through sample projects and literature review. Today, software companies offer many tutorial videos and trial times to test their systems. Some authors Pinho et al. (2014) studied the determinant factors in the adoption and recommendation of qualitative research software. They analysed five factors: (i) Difficulty of use; (ii) Learning difficulty; (iii) Relationship between quality and price; (iv) Contribution to research; and (v) Functionality. These authors indicate the two first factors as the most cited ones in the corpus of the data analysed:

- "NVivo is not exactly friendly. I took a whole course to learn to use it, and if you don't use it often enough, you're back to square one, as those "how-to" memories tend to fade quickly." **Difficulty of use**
- "I use Nvivo9 and agree that it is more user-friendly than earlier versions. I do not make full use of everything you can do with it however—and I've never come across anyone who does" "NVivo". **Learning difficulty**

In this context it is very important to study the User Experience and Usability of the QDAS, because these types of tools need to be at the service of the researcher, and not the opposite, therefore reducing the initial learning time and increasing the effectiveness and efficiency of all processes of analysis. Usability is an important dimension in the design and development of software. It is important to understand when to involve the user in the process.

3 Design Through Research: Usability Evaluation

Scientific research based on the Research and Development (R&D) methodology has a prevalence for quantitative studies. Primarily, it is in the interest of the researcher to test/prove some theory, through the actions of individuals involved in the study. The researcher intends to generalize and uses, usually, numerical data. When we apply an R&D to design of software packages, it becomes poor to reduce the researcher to someone who does not attempt to perceive the context in which the study takes place, interpreting the meanings of the participants resorting to interactive and iterative processes. Thus, the R&D methodology (or Design Research, also emerging under the expression Research through Design) has been gaining ground in software projects that, according to Pierce (2014), "include devices and systems that are technically and practically capable of being deployed in the field to study participants or end users" (p. 735). Zimmerman et al. (2007) state that Design Research "mean[s] an intention to produce knowledge and not the work to more immediately inform the development of a commercial product" (p. 494).

Associated to this methodology, expression such as Human-Computer Interaction (HCI), User-Centred Design (UCD) e Human-Centred Design (HCD) crop up. Maguire (2001), in his study "Methods to support human-centred design" tackles the importance of software packages being usable and how we can reach that wish. This study lists a series of methods that can be applied in the planning, comprehension of the context of use, in the definition of requirements, in the development and evaluation of project solutions.

According to Van et al. (2008), depending on the phase of the project, evaluation can serve different purposes. At the initial phase, when there still does not exist any software, evaluation provides information that supports decision making; at an intermediate phase and through the use of prototypes, it allows the detection of problems; at a final phase, and already in possession of a complete version of the software, it allows to assess quality. The process offered by these authors, named Iterative Design, is divided in four phases (Van et al. 2008):

- Without *software*: one aims at making decisions through collection of data with questionnaires, interviews and focus-groups and its analysis. These instruments are developed to characterise and define the user requirements;
- Low-Fidelity Prototype: one aims at detecting problems through gathering and analysis of data obtained from interviews and focus-groups. One identifies the need to feel appreciation, perceived usefulness and security and safety aspects;
- High-Fidelity Prototype: one aims at detecting problems through gathering and analysis of data obtained from questionnaires, interviews, think-aloud protocols and observation. Apart from the previous metrics, one aims at assessing comprehensibility, usability, adequacy, and the behaviour and performance of the user;
- Final version of the *software*: in the final version, the same data gathering instruments are applied. Besides some of the metrics mentioned before, one aims at perceiving the Experience and Satisfaction of the User.

According to Godoy (1995) "a phenomenon can be better understood in the context in which it occurs and of which it is part, having to be analysed from an integrative perspective" (p. 21)[1]. It is along this framework that in specific moments in the Iterative Design phases the researcher goes to the field to "capture" the phenomenon under study from the point of view of the users involved, considering

---

[1]Our translation.

every relevant point of view. The researcher collects and analyses various types of data to understand the dynamics of the phenomenon (Godoy 1995).

The recent need to identify and understand non-measurable/non-quantifiable aspects of the experience of the user is leading several researchers in the area of HCI to take hand of quantitative methods. HCI researchers started to understand that the context, both physical and social, in which specific actions take place along with associated behaviours, allow, through a structure of categories and their respective interpretation, to analyse a non-replicable phenomenon, not transferable nor applicable to other contexts (Costa et al. 2015).

The criteria defined by the standards are essentially oriented to technical issues. However, for a qualitative analysis to support that a software is of quality, it is necessary to take into account the research methodologies to be applied. Being an authoring software, researchers/users need to have knowledge of the techniques, processes and tools available in terms of data analysis in qualitative research.

4 WebQDA Usability Evaluation: Methodological Aspect

Despite the ISO 9126 providing 6 dimensions, in this study we will focus on the proposal to evaluate the usability of webQDA qualitative analysis software.

For an effective understanding of usability, there are quality factors that can be assessed through the evaluation criteria. Collecting and analysing data to answer the following questions will help determine if the software is usable or not (Seffah et al. 2008):

- Is it easy to understand the theme of the software? (Understandable)
- Is it easy to learn to use it? (Ease of learning)
- What is the speed of execution? (Use efficiency)
- Does the user show evidence of comfort and positive attitudes to its use? (Subjective satisfaction).

The answers to these questions can be carried out through a quantitative, qualitative or mixed methodology. However, we believe that they all have strengths and weaknesses that can be overcome when properly articulated.

Our study is based on the application of two questionnaires: (1) General use of the software, applied to the participants of the Iberian-American Congress on Qualitative Research—CIAIQ (years 2013, 2014 and 2015); and (2) review of the use of webQDA, applied only to users of this software. We obtained 98 replies in this second questionnaire. The survey questionnaire administered to CIAIQ participants obtained 362 answers. This questionnaire analysed only the answers to the question "When selecting software to support qualitative analysis, what criterion/criteria did you follow/would you follow for your choice?," related to Usability (see Fig. 1).

Fig. 1 Software usability
(Costa et al. 2016)

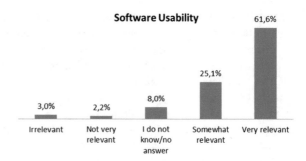

Of the participants, 86.7 % (n = 314) defined as selection criteria "somewhat relevant" qualitative analysis software (n = 91, 25.1 %), and "very relevant" (n = 223 61.6 %). "Do not know/" was chosen by 29 of the participants and 5.2 % chose "not very relevant" (n = 11) and "irrelevant" (n = 8). Focusing our analysis on software webQDA, version 2.0, we analysed the qualitative data that allow to supplement what was considered in the evaluation of its usability. When we used the System Usability Scale (SUS) (Brooke 1996) we got the average central tendency of 70 points (SD = 14.2), which allows us to conclude that webQDA 2.0 was "acceptable" in terms of usability, according to the SUS criteria (Costa et al.2016).

In this article we will examine the open questions of the survey applied to webQDA users (98 replies). Open questions were as follows: (i) What are the three main positive aspects of webQDA? (ii) What are the three main limitations of webQDA? (iii) What features would you propose for a new version of webQDA? (iv) Suggestions and general comments to help improve webQDA (training, references, etc.). To analyse the answers to these questions we used webQDA itself, under a simple analysis system as shown in Fig. 2.

In the analysis dimension "Features/Navigation" we tried to verify the positive and negative aspects of the various subroutines and webQDA features, as well as factors related to the navigability of the different screens. Table 1 shows some attributes of the users who responded to the survey applied in relation to the positive and negative aspects of this dimension. We found out, according to the

Fig. 2 Code dimension of
analysis from webQDA about
webQDA usability

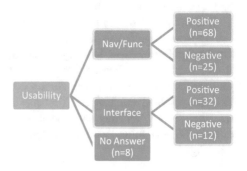

Table 1 Dimension functionality and navigability by attributes

Attributes	Positive (%)	Negative (%)	Did not answer (%)
Qualifications			
Graduation	0	0	0
Master in development	4	2	0
Master's degree	0	0	1
PhD development	21(55)	13(34)	4(10)
Completed PhD	43(77)	10(18)	3(5)
Research area			
Education and teaching	35(61)	18(31)	4(7)
Engineering and technology	5(55)	3(33)	1(11)
Health sciences	16(72)	4(40)	2(9)
Social sciences	12(92)	0	1(8)
Other	4	1	0
Professional situation			
Student	7(78)	1(11)	1(11)
Teacher	52(70)	18(24)	4(5)
Researcher	12(70)	3(17)	2(11)
Other	2(22)	5(55)	1(11)

"Qualifications" attribute, that users who had completed a PhD have a more positive perception (77 %) than those who are still developing their PhDs (55 %).

Some of these positive assessments may be perceived from the following statements:

- "Ability to join or cross two projects that work the same database, or similar database" (Q5, Female; 39; Brazilian; PhD completed; Health Sciences; Investigator/Researcher).
- "Work with internal and external sources in a simplified manner; ease of compiling qualitative data; quick access to graphics" (Q7, Female, 55; Portuguese; PhD in development; Other; Teacher).
- "Easy import of data; undo actions without harming the entire analysis" (Q98, Female; 41; Brazilian; Masters under development; Teaching and Education Sciences; Teacher).

In addition to these positive assessment on webQDA features one can find some negative comments like:

- "There should be a "help" in the interactive programme or even a more direct aid sector, via skype or another interactive environment ..." (Q10, Female; 58; Brazilian; PhD completed; Engineering and Technology; Teacher; Investigator/Researcher).
- "Platform frailties that is not available on tablet nor on mobile" (Q16, Male; 44; Brazilian; PhD under development; Teaching and Education Sciences; businessman).

Table 2 Dimension interface by attributes

Attributes	Positive (%)	Negative (%)	Did not answer (%)
Qualifications			
Graduation	0	0	0
Master in development	2(50)	2(50)	0
Master's degree	1(50)	0(0)	1(50)
PhD development	12(57)	5(24)	4(40)
Completed PhD	17(68)	5(20)	3 (12)
Research area			
Education and teaching	21(70)	5(17)	4(13)
Engineering and technology	2(50)	1(25)	1(25)
Health sciences	4(57)	1(14)	2(28)
Social sciences	4(57)	2(28)	1(14)
Other	3(43)	4(57)	0
Professional Situation			
Student	4(80)	0	1(20)
Teacher	24(67)	8(22)	4(11)
Researcher	6(50)	4(33)	2(16)
Other	0	0	1(100)

Table 2 shows the positive and negative assessment of the participants in relation to the webQDA interface. Here we cannot point out a difference between PhD students and those who have completed their PhDs. Some of the positive appreciations as to the Interface can be drawn from the following statements:

- "Although somewhat intuitive it requires mastery of analytical techniques; Operationalize more" (Q29, Female; 53; Portuguese; PhD under development; Education Sciences and Teaching; Teacher; Information organisation).
- "Better visibility of data. Better organization of the data which favours analysis" (Q40, Female; 47; Brazilian; PhD completed; Health Sciences; Teacher, Investigator/Researcher).
- "Being networked, easy viewing layout" (Q54, Female; 35; Brazilian; PhD completed; Health Sciences; Teacher; language).

Negative assessments about the interface:

- "Not all commands are intuitive: only with explanation is it possible to know its usefulness" (Q5, Female; 39; Brazilian; PhD completed; Health Sciences; Investigator/Researcher).
- "I think some menus could be more intuitive" (Q6, Male; 44; Portuguese; PhD under development; Other; Teacher).

Overall the perception of the webQDA interface is positive, although several aspects that should be improved were pointed out. Table 2 shows the positive and negative assessment of the participants in relation to the webQDA interface. Here we cannot point out a difference between PhD students and those who have completed their PhDs.

5 Final Remarks

Although no usability metric was referred in this article, qualitative analysis of the open questions in a questionnaire to users of the webQDA software showed positive and negative aspects that fall into two broad dimensions of analysis: (i) Functionalities/Navigation; and (ii) Interface. The main conclusions that we can draw from this analysis are:

- The users in general have a positive perception of the webQDA software and its features, namely Navigation and Interface.
- There seems to be, on average, a more positive assessment from users with completed PhDs as compared to those with their "in progress" doctoral programmes for the dimension Functionality/Navigation. This difference does not exist in the Interface dimension.
- There are no relevant differences between the views obtained from different professional or research areas, although a more positive assessment may be drawn as to Functionality/Navigation from users from the area of Education and Teaching (See Table 2).

Although most studies on usability define quantitative metrics, our study is a contribution that shows that qualitative analysis has great potential to deepen various dimensions of usability and functional and emotional interrelationships. Future studies should aim to diversify sources of data such as clinical interviews, semi-structured interviews and focus groups in advanced training contexts of qualitative research supported by webQDA.

Acknowledgments The first author thanks the Foundation for Science and Technology (FCT) the financial support that enabled the development of this study and presentation. The authors thank Micro IO company and its employees for the development of the new version of webQDA and participants of this study.

References

Cisneros-Puebla, C. A. (2003). Analisis cualitativo asistido por computadora. *Sociologias, 9*, 288–313.
Corti, L., & Gregory, A. (2011). CAQDAS comparability. what about CAQDAS data exchange? *forum: Qualitative. Social Research, 12*(1), 1–17.
Costa, A. P., et al. (2016). webQDA—Qualitative data analysis software usability assessment. In: *Conferência Ibérica de Sistemas e Tecnologias de Informação*. Gran Canária—Espanha: AISTI—Associação Ibérica de Sistemas e Tecnologias de Informação, (in press).

Costa, A.P., Faria, B.M. & Reis, L.P. (2015). Investigação através do Desenvolvimento: Quando as Palavras "Contam." *RISTI—Revista Ibérica de Sistemas e Tecnologias de Informação*, (E4), pp.vii–x. Available at: Investiga??o atrav?s do Desenvolvimento: Quando as.

Godoy, A. S. (1995). Pesquisa Qualitativa: tipos fundamentais. *Revista de Administração de Empresas, 35*(3), 20–29.

ISO9241-210. (2010). Ergonomics of human-system interaction (210: Human-centred design for interactive systems).

Kaefer, F., Roper, J. & Sinha, P. (2015). A software-assisted qualitative content analysis of news articles : Example and reflections. *Forum Qualitative Sozialforschung, 16*(2).

Maguire, M., (2001). Methods to support human-centred design. *International Journal of Human-Computer Studies, 55*(4), 587–634. Available at: http://linkinghub.elsevier.com/retrieve/pii/S1071581901905038 Accessed November 1, 2013.

Neri de Souza, F., Costa, A.P. & Moreira, A. (2011). Questionamento no Processo de Análise de Dados Qualitativos com apoio do software webQDA. *EduSer: revista de educação, Inovação em Educação com TIC, 3*(1), 19–30.

Pierce, J. (2014). On the presentation and production of design research artifacts in HCI. In: *Proceedings of the 2014 conference on Designing interactive systems—DIS'14*. New York, New York, USA: ACM Press, pp. 735–744. Available at: http://dl.acm.org/citation.cfm?doid=2598510.2598525.

Pinho, I., et al. (2014). Determinantes na Adoção e Recomendação de Software de Investigação Qualitativa: Estudo Exploratório. *Internet Latent Corpus Journal, 4*(2), 91–102.

Puebla, C.A.C. & Davidson, J. (2012). Qualitative computing and qualitative research: Addressing the challenges of technology and globalization. *Historical Social Research, 37*(4), 237–248. Available at: http://www.jstor.org/stable/41756484.

Richards, L. (2002a). Qualitative computing–a methods revolution? *International Journal of Social Research Methodology, 5*(3), 263–276.

Richards, L. (2002b). Rigorous, rapid, reliable and qualitative? computing in qualitative method. *American Journal Health Behaviour, 26*(6), 425–430.

Scanlon, E., Mcandrew, P., & Shea, T. O. (2015). Designing for educational technology to enhance the experience of learners in distance education: How open educational resources, learning design and moocs are influencing learning. *Journal of Interactive Media in Education, 1*(6), 1–9.

Seffah, A., et al. (2008). Reconciling usability and interactive system architecture using patterns. *Journal of Systems and Software, 81*(11), 1845–1852. Available at: http://linkinghub.elsevier.com/retrieve/pii/S016412120800085X Accessed November 1, 2013.

Souza, F. N., Costa, A. P., & Moreira, A. (2011). Análise de Dados Qualitativos Suportada pelo Software webQDA. *VII Conferência Internacional de TIC na Educação: Perspetivas de Inovação* (pp. 49–56). VII Conferência Internacional de TIC na Educação: Braga.

Souza, F.N. de, Costa, A.P. & Moreira, A. (2016). webQDA. Available at: www.webqda.net.

Teixeira, R. A. G., Neri de Souza, F., & Vieira, R. M. (2015). *Docentes Investigadores de Programas de Pós-graduação em Educação no Brasil: Estudo Sobre o Uso de Recursos Informaticos no Processo de Pesquisa* (pp. 741–768). No prelo (: Revista da Avaliação da Educação Superior.

Tesch, R. (1991). Introduction. *Qualitative Sociology, 14*(3), 225–243.

Van Velsen, L. et al. (2008). User-centered evaluation of adaptive and adaptable systems: A literature review. *The Knowledge Engineering Review, 23*(03), 261–281. Available at: http://www.journals.cambridge.org/abstract_S0269888908001379.

Zimmerman, J., Forlizzi, J. & Evenson, S. (2007). Research through design as a method for interaction design research in HCI. In *Proceedings of the SIGCHI conference on Human factors in computing systems—CHI'07*. New York, New York, USA: ACM Press, p. 493. Available at: http://portal.acm.org/citation.cfm?doid=1240624.1240704.

Understanding Participatory Policymaking Processes: Discourse Analysis in Psychosociological Action Research

Roberto Falanga

Abstract Worldwide participatory policymaking with civil society has become a case in point for new patterns of governance at different levels. The enhancement of public services' quality compels intertwined challenges to participants, be they citizens, politicians, or civil servants. Given the limited attention that scientific literature has paid to understanding how roles and functions of civil servants are demanded to change through participatory processes, an exploratory action research was run in 2012 with 29 civil servants of the Municipality of Lisbon. Civil servants were all engaged in—at least—one of the four participatory processes run by the administration at that time: Participatory Budget; Bip/Zip; Local Agenda 21; Simplis. The action research was aimed at grasping the psychological change-driven dynamics played between these subjects, with the support of psychosociological theories and methods. Focusing on the analysis of the discourse, as one of the methods employed towards this end, this text focuses on the specific methodological apparatus of the discourse analysis approach. This contribution will hopefully open to further studies on discourse analysis in action research, and enhance the overall debate on the employment of qualitative methods in participatory policymaking studies.

Keywords Participation · Civil servants · Action research · Discourse analysis · Psycho-sociology

1 Introduction

Participatory policymaking processes with civil society were first experienced to reform political systems in Latin America, in the end of 1980s (Avritzer and Navarro 2003; Sousa Santos 2003). The perception of enlarged participation of civil society in decision-making aimed to effectively unfold social inequalities and economic redistribution. Inspired by some first successful experiences, several

R. Falanga (✉)
Institute of Social Sciences, University of Lisbon, Lisbon, Portugal

© Springer International Publishing Switzerland 2017
A.P. Costa et al. (eds.), *Computer Supported Qualitative Research*,
Studies in Systems, Decision and Control 71,
DOI 10.1007/978-3-319-43271-7_2

13

countries in Europe have assumed in the last two decades the opportunity to make participation a device for—inter alia—citizenry trust recovery, less electoral abstention, and new solutions before increasing complexity of transnational and multi-scale economic, financial and political networks. Saying so, the diffusion of participatory mechanisms cannot be understood by merely looking at political intentions. Participation also compels understanding about inherent organizational challenges for policymaking changes.

The reconfiguration of democratic goals and governance instruments, aimed to pursue effective policies, has largely challenged the bureaucratic rationale of public administration functioning (Denhardt and Denhardt 2007). Participatory processes demand the (re)organization of administrative levels, systems and connections in order to sustain the effective impact of policies. Together with the more "structural" dimension, the engagement of civil servants in such processes creates new forms of interactions that need to be further analyzed. Civil servants' symbolical representations and acting strategies construct new change-driven "semantics" of governance from within the organizational context (Falanga 2013).

The question is crucial, in as much as one of the critical points of participatory policymaking hitherto has entailed the narrow application of bureaucratic logics to the problems affecting society. The segmentation of administrative responses has often carried to "sectorialized" policymaking processes, less effective in a growingly complex scenario. As such, one of the most challenging effort of participatory processes, and more broadly of new governance models, has been that of promoting new integrated visions of social problems (Peters and Pierre 2007).

We decided to explore this topic by looking at the change from within: how do civil servants symbolically represent participation in a changing organizational context? The action research run in 2012 was based on a qualitative approach which included the discourse analysis of 29 individual interviews with the civil servants engaged in four participatory processes, run at that time by the Municipality of Lisbon. The text pretends to synthetically frame the qualitative approach to the case study and put the emphasis on the discourse analysis method. Towards the aim we will not provide deep information about the outcomes of the action research, which are retrievable in other publications of the author of this text Falanga (2013, 2014).

2 Making Sense the of Action Research

The "generative" function of the action research implies helping the subjects involved in the research to reflect and self-reflect on specific occurring changes, and possibly act within them. Such a perspective is consistent with Kurt Lewin's proposal, when referring to action research as both an instrument to better understand the changes going on in a system, and a way to perturb the system itself (Lewin 1948). The design of an action research should then be pursued according to the characteristics of specific contexts and be set in dialogical connection with the subjects, rather than provide a predefined set of methodological tools to be applied to "objects" of analysis (Scott 1965). Moreover, the action research should identify the set of scientific issues to be explored throughout the field-study itself. The identification should be open enough to conceive areas of research where the recognition of main theories should match provisory relations with progressive findings (Pagés et al. 1998).

Due to these appealing characteristics, social and political sciences have extensively debated about the adoption of new qualitative methods in the past few decades. In this vein, some scholars from the public policy analysis field have had a pervasive reaction to neo-positivist approaches Yanow (2000) and Fisher (2003) has clearly stated that:

> [...] the interpretive orientation on meaning requires the social scientist to pursue the unobservable as well. Because language is able to carry and transmit meanings among people, access to the realm of meaning often can be gained through the study of communication, both spoken and written. But such meanings are generally only indirectly made available through such communications. Thus it is necessary for the analyst to move beyond empirical methods – such as content analysis – to an interpretive reconstruction of the situational logic of social action (ibidem, pp. 139-140).

Inspired by this statement, we wonder how can we make sense of the logic(s) that construct communication? Narrative-in-making permits to better understand the ways policies can comprise a sequence of ambiguous claims (Czarniawska-Joerges 1997). However, narratives are neither all the same nor serve the same purpose. Storytelling, for instance, is likely to grasp a variety of information and thoughts by weaving them into a plot-making sense of complex situations (Bruner 1986). Kykyri et al. (2010) argue the central role of daily conversations in producing and managing organizational changes. Heracleous and Marshak (2004) further state that "organisational discourse analysis is not simply an intellectual luxury but can have pragmatic, relevant implications" (ibidem, p. 113). Likewise, Argyris (1991) makes an argumentative link between thought and action in opposition to the sharp differentiation between scientific and useable knowledge. In line with this, the author Argyris (1994) highlights the importance of enabling subjects with emerging knowledge, as a potential factor of change when considering the role of feelings. "Progress toward change requires expressing those feelings as well as respecting them. It is important for organisational participants to explore the reasons for their feelings" (ibidem, p. 353). The author further points out the critical task of interrupting the "closed cycle" of defensive reasoning and behavior in order to learn from and reflect on the meanings of change (ibidem).

Understanding discourse-making is key to intervene over social relationships and contexts. In these terms, rather than approaching change as radical shifting from one stable state to another, discourse-oriented methods applied within an action research can have the potential to both grasp and generate change.

Acknowledging the role of psychology in understanding symbolical representation and signification of the reality, as well as intervening towards goals of change, psychosociological approaches have especially contributed to new discourses and actions' analysis models (Carli and Paniccia 2003). These models have argued—alike the action research—the scientific necessity to progressively monitor and prove scientific outcomes concerning symbolical representations through the reiteration of interactive processes with the involved subjects (Carli and Paniccia 2003; Olivetti Manoukian 2007).

3   Case-Studying the Municipality of Lisbon

Case-studies can increase knowledge around a specific scientific issue and/or improve a scientific theory. In this sense, there can be either an explorative or descriptive purpose when choosing the case-study. In the first case, researchers can undertake pivotal studies to test new hypotheses with no strict demands on statistical relevance of their study samples. In the second case, neither predictions nor prescriptions are given, rather comprehensive descriptions of the case (Yin 2003).

The explorative purpose of our case-study was to understand the ways civil servants engaged with four participatory processes in the Municipality of Lisbon were symbolically representing participation within a driven-change organizational context (Falanga 2013). The four analyzed processes were: Participatory Budgeting (participatory allocation of a predefined public pot of money); Bip/Zip (financial and logistic support to local partnership-making for community-based intervention in priority areas); Local Agenda 21 (participatory consultation concerning urban sustainable goals); Simplis (participatory consultation concerning bureaucratic simplification of the local administration).

Regarding the transformative nature of participatory policymaking processes, we identified four intertwining organizational factors: (i) vertical lines of rules (internal hierarchy and distribution of roles between top-level, middle-level and street-level bureaucrats); (ii) Horizontal networks (intra-administrative relationships between former and new administrative units); (iii) Back-office functions (formulation and articulation of policy innovations at the institutional level); (iv) Frontline interaction with civil society (formulation and articulation of policy innovations at the "street" level) (Fig. 1).

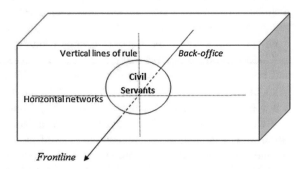

Fig. 1 (Civil servants producing change, figure of the author of this text)

The four identified factors highlight the structural dimension of the changes that civil servants are demanded to approach somehow in their daily organizational life. Considering the transformative nature of participatory processes, civil servants cannot help but being at the heart of our attention on change-driven symbolical representations of participation. Furthermore, the variability of these representations can be understood by looking at the ways they are shared between the civil servants, rather than approaching their analysis as incommensurably idiosyncratic to individuals.

The design of the action research was a participatory step of our approach. We negotiated the meaning of the proposed methods and goal with political representatives, administrative managers, and the 29 civil servants engaged with the four participatory processes. The resulting methodological plan was composed of five steps and was run between January and December 2012 (and adapted to each one of the four participatory processes' scheduling):

1. Understanding the context, focusing on local administration and the administrative teams demanded to manage participatory initiatives;
2. Collecting quantitative data regarding the four participatory processes in 2012, such as budgetary incomes and outputs, people involved, etc.;
3. Observing on the field the implementation of the four participatory processes, as regards both internal meetings between civil servants and assemblies with citizens;
4. Interviewing the 29 identified civil servants engaged with the processes in order to analyze their discourse about participation;
5. Planning regular follow-ups with the civil servants involved in the action research.

We will focus on the fourth step in the next section. Before that, it is worth underlining that the action research has undertaken systematic self-reflection on the observed relationships and shared reflection on emerging evidence with the involved subject. This psychosociological task is based on the psychoanalytical "contra-transfer" function in that it compels self-observing, reflecting on, and making sense of the emotions generated by and through our relationships. This task led us to formulate two main questions which introduce the discourse analysis. Firstly, interviews helped to make the interviewees feel they were the owners of key knowledge about participation. Such acknowledgment accomplishes the symbolical legitimization purpose of the civil servants' "experiential knowledge". Secondly, interviews helped to make interviewees aware about their key contribution to participatory studies. Such awareness can be considered as the result of the scientific legitimization purpose of the action research.

4 Focusing on the Discourse Analysis of Civil Servants

The analysis of the discourse can be approached from different theoretical and methodological frameworks. Some frameworks are aimed to understanding the common language adopted by a community (e.g. ethno-methodology concerning the social knowledge of subjects); others the illustration of the content by either decomposing the text into micro-units or analyzing its passages concerning the theme of investigation; some others the interpersonal construction of a theory, by matching sociolinguistic and rhetorical components (Pagés et al. 1998). The psychosociological perspective on communication concerns the ways it is produced as an action. Such an assumption compels understanding the interplay between unconscious and conscious sets of mind behind the production of discourses (Fornari 1979).

By adopting the psychosociological theoretical perspective, we established the methodological tools through which we wanted to address the shared symbolical dimensions of driven-change discourses (Falanga 2014). Towards this end we decided to cluster the different patterns characterizing the common symbolical object "participation" by following Carli and Paniccia's guidelines (2002). We proposed semi-structured interviews to each one of the 29 civil servants. Planned as one-hour interviews, interviewees variably employed from 40 min up to 1 h and 40 min to develop their discourses concerning participation. The initial question of the interview was: "What do you think about participation?." The purpose of such open and broad question was to give support to emotional self-exploration of the issue. Such goal is consistent with the exploratory purpose of the action research. Interviews have been conducted so as to let emotions and thoughts flow without a pre-determined script, even when it was perceived as confusing or too demanding (Goffman 1988). On basing their answer by making reference to both work and life experiences, civil servants were introduced to a profound and unexpected experience of "sense making" (Weick 1997).

The progressive engagement of civil servants to the interview was characterized by some common elements:

1. At the beginning, all interviewees felt fairly constrained in answering the question. The initial impasse was interpreted as a form to express two typologies of psychological "conflict":

   a. The question is planned to generate some anxiety which stems from its character of openness supposed to let the narrative flow into the topic.
   b. The question is received as a surprising demonstration of interest when not feeling they have often been asked about their vision of the tasks.

2. In the course of the process, we adopted different strategies in order to facilitate interviewees' narratives, and the interviewees themselves have generally taken possession of their arguments as time went on. Initial hesitancy has in almost all cases "thawed" into curiosity, possibly stemming from:

    a. Progressive word of mouth among colleagues about the occasion of being spokesmen and spokeswomen of their own job experiences.

    b. Progressive trust in the researcher due to the encouraging emotional tone of the interview aimed at neither "invading" interviewee with lots of questions, nor evaluating his/her job performance.

3. At the end of the interviews, three general recurring items have emerged at the content-level:

    a. The variety of participatory processes implemented by the Municipality of Lisbon, possibly relating to their overall context of belonging.

    b. The participatory processes as phenomena impacting both internal and external actors, possibly referring to their direct experience with participation.

    c. The significance of participation in terms of one's own career, possibly referring to their investment in terms of a lifelong career.

Once we interviewed the civil servants, we proceeded to the analysis. The analysis of the discourse aimed at catching the deep emotional and psychological dynamics behind and beyond discourse production by deconstructing the intentional connections among the words (Carli and Paniccia 2002, 2003; Battisti and Dolcetti 2012). The analysis of the discourse was finally complemented by observation and data collection (Carli 2006).

Towards the aim, we first created a sole corpus of transcribed text and identified five independent variables aimed at characterizing the results related to our case-study. The function of independent variables is to show their relevance in the formation of clusters, i.e. after the clusterization is run. The variables were: (1) sex; (2) function; (3) participatory process; (4) years of work for the Municipality; (5) years of experience in participation. It is worthy to clarify that the fourth variable served to specify possible differences resulting from the status of lifelong career that all the civil servants taking part to the action research were carrying on. Also, the fifth variable was conceived as a way to differentiate years of engagement in participatory processes and see possible effects on the process of clusterization.

The sole corpus of text was secondly processed through the text analysis software Alceste, standing for "Analysis of Lexemes Co-occurring within Simply Textual Enunciations" (Reinert 1995). This software works through segmenting the corpus of text into Unities of Elementary Context (UCE), which are statements or paragraphs, and the definition of categories of words with their lexemes. After obtaining the complete set of words present in the corpus of text, we checked the whole vocabulary in order to clean it from non-relevant words and keep the key ones. The elimination of the non-relevant words implied defining criteria consistent with the whole design of the action research. Moreover, we also eliminated words holding pure syntactical functions, such as articles, prepositions, etc. Finally, we eliminated the words sharing the lexeme referring to "participation", for it was used as focus of the question. Indeed, the invitation to talk about participation implies considering the speech originated by this stimulus—rather than including the stimulus itself—and identifying multiple ways of perceiving it.

As regards the selection of keywords, we created lexemes aimed at joining words referring to the same headwords (e.g. "to serve", "service", "servant" merged into one headword: "serv_"). In doing so, we also identified ambiguous acceptations of some words in the written forms or in the cultural use, as well as detected words that when joined together express precise concepts (e.g. the two concurrent words "sem abrigo" transformed into one "sem_abrigo", meaning "homeless"). Finally, we checked the emotional "density" of the words in order to make a final decision on the relevant ones.

This preparatory phase was necessary to work two statistical operations out through a second use of the software Alceste: multiple correspondences and cluster analysis of the keywords. The clusterization of the UCE into groups of co-occurring keywords listed the frequency of words in relation to one another within UCE (multiple correspondences) as well as proximity and distance between UCE (cluster analysis).

According to our psychosociological approach, the clusters enlighten about different modalities through which the subjects symbolically represent the object of analysis, in this case participation. The crossed detection of multiple correspondences among the UCE and the clusterization of the co-occurring words within the text respond to phenomena related to "free associations" outputs derived from the evocative stimulus provided by the question (Carli and Paniccia 2002).

Four clusters emerged by the discourse analysis. The clusters neither match the discourse of individuals or groups nor identify specific participatory processes. It is a mere coincidence that the outcome was of four clusters for four participatory processes. The clusters characterize cross-cutting psychological dimensions at the heart of civil servants' symbolical representations of participation. The spatial organization of the clusters within one factorial space and their relationship to the factorial axes represents a further source of information on the inner diversity of representations (Fig. 2).

At the statistical level, the relation among clusters is read by taking into account the distance from the factorial axes (Table 1). In psychosociological terms it means considering axes as social factors which condense the key instances emerging from the clusters.

The UCE have different statistical weight in each Cluster, what means that the frequency of co-occurring keywords is variably distributed within the clusters (Table 2). In psychological terms, it means considering how specific symbolical dynamics prevail in each clusters, in detriment of others.

As regards the interpretation of the co-occurring keywords in each cluster, we first made reference to the etymology of each keyword, as a way to provide our analysis with an historical and cultural background on the use of Portuguese language. We developed a deep reflection concerning the resources and limits of cultural translation that characterized the whole process of the action research, which was especially evident in this phase due to our linguistic limits and resources. The multiple levels of translation including the different steps of the research represented a highly challenging issue that we sought to experience by guaranteeing

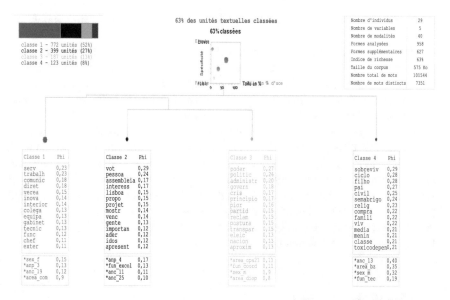

Fig. 2 (Factorial space, output of the discourse analysis run by the author of this text)

Table 1 Factorial axes and clusters

First factorial axis	Polarization between the Cluster I (+0.314) and the Cluster IV (−1.532)
Second factorial axis	Polarization between the group of Clusters I/IV (−0.313/−0.600) and the group of Clusters II/III (+0.571/+0.534)
Third factorial axis	Polarization between the Cluster II (−0.414) and the Cluster III (+0.917)

Table 2 Clusters and classified UCE

Cluster	Classified UCE (%)
I	772 (52)
II	399 (27)
III	187 (13)
IV	123 (8)

the quality of the outcomes. Said so, we took benefit from plural etymological sources to better reduce the psychological polysemy of the emerging keywords.

Semantic polysemy corresponds to infinity sets of meanings that the unconscious state of mind produces underneath our understandings. Polysemy is not a "mere" multiplicity of meanings, it is rather the simultaneous "sense making" of our rational understanding of the world (Matte Blanco 2000). Said so, our interpretation involved focusing not only on the multiple intended meanings of the keywords, but rather looking at their co-occurrence in order to grasp emerging emotional dynamics.

The purpose was that of providing the subjects with access to the symbolical sense at the heart of their own discourses, by revealing the shared dimensions beyond their individual experience with participatory processes. The impact of this type of knowledge is evident on the improvement of both individual and social awareness, as well as on the potential for action.

Being the purpose of the action research expressly exploratory, the external validity of this discourse analysis was ensured by: (i) the triangulation within a set of qualitative methods—data collection and observation—both converging and improving the discourse analysis' results; (ii) the positive feedback of the civil servants in the regular follow-ups, who affirmed to assume the emerging cultural patterns as their own expression. Finally, the internal validity of this method has been comprehensively proved by, inter alia, Carli and Paniccia (2002), Battisti and Dolcetti (2012), and Falanga (2013).

5 Conclusive Reflections

Participation of civil society in policymaking processes encourages the adoption of a complex set of changes within organizational contexts. Along with structural changes, new governance equilibriums generate ambivalent emotions, desires and fears by the involved subjects, be they politicians, citizens, or civil servants. Supported by psychosociological theories and methods, our framework intended to give voice to what is often kept unexpressed—or what is looking for words to be expressed—in participatory processes. The generation of different symbolical representations of change and the observed impact of these representations on the daily performance with policymaking processes, compelled us to explore the ways civil servants represent participation.

Towards the aim, we focused on four participatory processes developed by the Municipality of Lisbon in 2012. Our action research included the adoption of a discourse analysis method, which was applied along with 29 civil servants engaged in the four processes. Our goal was to explore their profound experience with participation. From the analysis of four clusters of co-occurring keywords we picked out four different patterns of symbolical representation. By sharing the outcomes with the civil servants, we not only negotiated new meanings, but also and mainly set specific setting for shared reflection on the ways change can be produced, improved, or contrasted. In doing so, we elicited a shared self-reflexive setting with civil servants who had the opportunity to get aware of "new" meanings about their profound connection with the participatory processes, and also to inspire new steps for action research.

References

Argyris, C. (1991). The use of knowledge as a test for theory: The case of public administration. *Journal of Public Administration Research and Theory, 1*(3), 337–354.

Argyris, C. (1994). Initiating change that perseveres. *Journal of Public Administration Research and Theory, 4*(3), 343–355.

Avritzer, L., & Navarro, Z. (Eds.). (2003). *A inovação democrática no Brasil*. São Paulo: Cortez.

Battisti, N., & Dolcetti, F. (2012). Emozioni e testo: Costruzione di risorse per il tagging automatico. In: *Paper presented to the 11es Journées internationales d'Analyse statistique des Données Textuelles*. Paris, France.

Bruner, J. (1986). *Actual minds, possible worlds*. Cambridge: Harvard University Press.

Carli, R. (Ed.). (2006). *La scuola e i suoi studenti: un rapporto non scontato/L'école et ses élèves: des rapports à ne pas tenir pour acquis/La escuela y sus estudiantes: una relación que no se da por descontado*. Milan: Franco Angeli.

Carli, R., & Paniccia, R. (2002). *L'analisi emozionale del testo. Uno strumento psicologico per leggere testi e discorsi*. Rome: Franco Angeli.

Carli, R., & Paniccia, R. (2003). *L'analisi della domanda. Teoria e tecnica dell'intervento in psicologia clinica*. Bologna: Il Mulino.

Czarniawska-Joerges, B. (1997). *Narrating the Organization: Dramas of Institutional Identity*. Chicago: The University of Chicago Press.

Denhardt, R., & Denhardt, J. (2007). *The New Public Service: Serving, not Steering*. New York/London: M.E. Sharpe.

Falanga, R. (2013). *Developing change. A Psychosociological action research with civil servants engaged in participatory processes*. Doctoral Thesis, University of Coimbra.

Falanga, R. (2014). Changes need change: A psychosociological perspective on participation and social inclusion. *Rivista di Psicologia Clinica, 2*, 24–38.

Fischer, F. (2003). *Reframing Public Policy. Discursive Politics and Deliberative Practices*. New York: Oxford University Press.

Fornari, F. (1979). *I fondamenti di una teoria psicanalitica del linguaggio*. Turin: Boringhieri.

Goffman, E. (1988). *Il rituale dell'interazione*, tr.it *Interaction Ritual* (1967). Bologna: Il Mulino.

Heraclous, L., & Marshak, R. (2004). Conceptualizing organizational discourse as situated symbolic action. *Human Relations, 57*(10), 1285–1312.

Matte Blanco, I. (2000). *L'inconscio come insiemi infiniti: Saggio sulla bi-logica*, tr.it *The Unconscious as Infinite Sets: An Essay in Bi-Logic* (1975). Turin: Einaudi.

Olivetti Manoukian, F. (2007). Formazione e Organizzazione. *Spunti, 10*, 107–144.

Kykyri, V., Puutio, R., & Wahlstrom, J. (2010). Inviting participation in organizational change through ownership talk. *Journal of Applied Behavioral Science, 46*(1), 92–118.

Lewin, K. (1948). *Resolving Social Conflicts: Selected papers on group dynamics*. New York: Harper & Brothers.

Lipsky, M. (1980). *Street-level Bureaucracy: Dilemmas of the individual in Public Services*. New York: Russel Sage Foundation.

Pagés, M., De Gaulejac, V., Bonetti, M., & Descendre, D. (1998). *L'emprise de l'organisation*. Paris: Presses Universitaires de France.

Peters, G., & Pierre, J. (2007) Governance and civil service systems: from easy answers to hard questions In: C.N. Jos Raadschelders A.J. Theo Toonen; M. Frits Van der Meer (Eds.). *The Civil Service in the 21st Century. Comparative Perspectives*. New York: Palgrave MacMillan.

Reinert, M. (1995). I mondi lessicali in un corpus di 304 racconti di incubi attraverso il metodo Alceste In: Roberto Cipriani; Sergio Bolasco (Eds.). *Ricerca qualitativa e computer*. Milan: Franco Angeli.

Sousa Santos, B. (Ed.). (2003). *Democratizar a democracia: os caminhos da democracia participativa*. Porto: Edições Afrontamento.

Yanow, D. (2000). *Conducting interpretive policy analysis*. London: Sage.
Yin, R. (2003). *Applications of Case Study Research*. Thousand Oaks, CA: Sage.
Weick, K. (1997). *Senso e significato dell'organizzazione*, tr.it *Sensemaking in Organizations* (1995). Milan: Raffaello Cortina Editore.

Combined Use of Software that Supports Research and Qualitative Data Analysis: Potential Applications for Researches in Education

Katia Alexandra de Godoi e Silva
and Maria Elizabeth Bianconcini de Almeida

Abstract This article analyzes the potential applications of the articulation between software programs designed to investigate and analyze qualitative data similarities on researches in the education field. The first part of this article discusses the context and the meta-ethnographic process. Next, it explores the use and articulation of software programs designed to support qualitative research, namely webQDA and CHIC. The third part presents the weaving of the meta-ethnographic analysis. The final remarks focus on the integration of this study's findings, synthesis and conceptualization, combined with the acknowledgment of the contributions derived from the articulated use of both programs to develop this research.

Keywords webQDA software · CHIC software · Qualitative analysis · Research in education · Meta-ethnographic approach

1 Introduction

This study analyzes the potential applications of the articulation between software programs designed to investigate and analyze qualitative data similarities in research work in the education field.

We focus on two specific software programs: the web Qualitative Data Analysis (webQDA)—a qualitative data analytic tool in a collaborative online environment —and the Correspondence and Hierarchical Cluster (CHIC).

K.A. de Godoi e Silva (✉)
Postgraduate Program in Education, Catholic University Dom Bosco,
Dom Bosco, Brazil
e-mail: katigodoi@gmail.com; 3085@ucdb.br

M.E.B. de Almeida
Postgraduate Program in Education and Curriculum, Pontifical Catholic
University of São Paulo, São Paulo, Brazil
e-mail: bbethalmeida@gmail.com; bethalmeida@pucsp.br

© Springer International Publishing Switzerland 2017   25
A.P. Costa et al. (eds.), *Computer Supported Qualitative Research*,
Studies in Systems, Decision and Control 71,
DOI 10.1007/978-3-319-43271-7_3

The relevance of the present study lies in the potential incorporation of these two programs combined in a research where the treatment of data using the first software (webQDA) allows for organizing them in categories and/or themes, hence preparing them to be used by the second one (CHIC). Such strategy enabled the researchers to retrieve the processing performed with both software programs at different moments of their work as a means to search for new significant elements in the ongoing study and, also, to deepen the analyses.

In order to explain how we developed the articulation between these two software programs, we synthesized the results of a study conducted within a research (Almeida 2013; Godoi 2013), using the meta-ethnographic methodology. For this, we present the context and the meta-ethnographic process next. After that, we discuss the improvements in the use of software tools in qualitative research. Finally, we offer the analysis of the findings resulting from this study.

2   The Context and the Meta-Ethnographic Process

This work chose, as its investigative object, the issue of synthesizing results from a study conducted within a project (Almeida 2013), funded by the grant of "One Computer per Student Program" (PROUCA 2010) of the Brazilian Agency for Scientific and Technological Development (CNPq), the Coordination for the Enhancement of Higher Education Levels (Capes) and the Ministry of Education (MEC). This study is delimited by a teacher development course held in an elementary school that participates of the PROUCA program, located in a municipality in the state of São Paulo, Brazil.

Project and Program UCA were inspired by the Project One Laptop per Child (OLPC) designed by Nicolas Negroponte, whose ideas were launched in Brazil in 2006 when the one laptop per student option was adopted by the public schools. The recipient schools were in charge of managing the equipment distribution to students, and this feature sometimes restrained the use of laptops solely to activities developed inside the school. In order to provide the professional development for teachers so that they could build their basic technological skills as well as help them with the integration of technology and curriculum as proposed by Shapley et al. (2010), and Suhr et al. (2010), the development course uses the b-learning[1] modality. In the selected school, the development program was a direct responsibility of the Pontifical Catholic University of São Paulo, a private university in the state of São Paulo, Brazil.

---

[1]B-Learning, or Blended Learning, is considered a hybrid learning modality, that is, the teaching-learning processes involve both on-site and online contexts, and they can resort to different methodologies and resources.

Fig. 1 A simple model
relating learner, content, and
context in a learning event
(Quoted from Figueiredo and
Afonso 2006, p. 5)

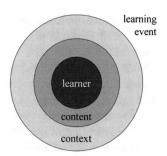

This type of development using b-learning modality involved theory-practice workshops and online activities in the learning environment e-Proinfo,[2] on the choice and evaluation of digital learning material (DLM) (Godoi and Padovani 2009; Godoi 2013; Nokelainen 2006; Squires and Preece 1996, 1999), having as a basis the Contextual Evaluation Plan (Ramos et al. 2004), which allowed educators to reflect upon their choices before, during and after the use of DLM.

The concept of context is based on Figueiredo and Afonso (2006), who bring relevant contributions for the understanding of what the term "context" means. The authors adopted a simplified model that relates the learner (e.g., the in-service teacher) to the content and the context in a learning event (e.g., the selection of digital teaching material). This model proposes three definitions: (1) A learning event is a situation where an individual learns; (2) The content is the information that has been structured and coded as a text (e.g., the Contextual Evaluation Plan; the developer's spoken words or any other media used in on-site or online development); (3) Context is the set of relevant circumstances for the learner (in this case, the teacher) to build his knowledge (Fig. 1).

In this model, presented in Fig. 1, we can see that the lines between the layers—content and context—become blurred, that is, from the content layer you can cross into the context layer and vice versa. In this view, "one generates the other" and "one cannot exist without the other", which is a feature that is shared by the development course under analysis in this article.

This cross section is restricted to two teachers who contributed with their reflections, interacted in the e-Proinfo forums and produced written reports using the Contextual Development Plan. The individuals are identified according to the code consisting of a number followed by letters and one numeral referring to the mnemonic indicating their teaching level and/or position they hold, such as: EST1 = Elementary School Teacher I,[3] EST2 = Elementary School Teacher II.[4]

---

[2]e-Proinfo. Retrieved April 30, 2016, from e-proinfo.mec.gov.br.

[3]The Brazilian elementary school I (1st through 5th grade) is equivalent to the United States elementary school level (1st through 5th grade).

[4]The Brazilian elementary school II (6th through 9th grade) is equivalent to the United States middle school level (6th through 8th grade).

We chose meta-ethnography for the methodological approach, according to Noblit and Hare (1988) and Alarcão (2015), as a process of meticulous analysis founded on an interpretive approach that involves data re-interpretation besides favoring a contextualized analysis. The meta-ethnographic researcher's role is to interpret, re-configure, re-construct and synthesize in the sense that he reinterprets and evaluates the scope of findings in qualitative research. In meta-ethnography— besides data synthesis—we often perform analysis, integration, transformation and conceptualization of qualitative results.

We used the meta-ethnographic approach for the analysis and synthesis of findings from a research previously conducted by Godoi (2013). The authors of this study used such findings to obtain the themes. The process was conducted following the steps below, as proposed by Noblit and Hare (1988), and re-signified by the authors based on the characteristics of the research data:

- Re-read the study, take notes, reformulate and create themes and key sub-themes;
- Determine how the themes are related;
- Synthesize the findings, work on integrating them and identify the changes;
- Write out the synthesis and conceptualize the results.

For this process to be rigorous and clear, the researcher needs to recover the context of the original study and reinterpret the results without ignoring the original author's viewpoint (Alarcão 2015).

It is worth underscoring that the meta-ethnographic approach does not imply undervaluing the methodology used originally, namely the Design-based Research (PBD) (Reeves 2000), since meta-ethnography is characterized by its flexibility to incorporate aspects that emerged from the situation under study.

3   Software Used to Support Qualitative Research: The Combined Use of webQDA and CHIC

In the last 30 years, there has been a growing development and interest for qualitative research approaches that are able to incorporate the specific and complex features of contexts under study and, at the same time, improve scientific rigor. Groups of researchers started to focus on software creation, many of which became accessible products available for the academic community, some with specific licenses and some license-free. Since then, the use of software tools in scientific investigations of qualitative or qualitative-quantitative bases has gained important recognition in the scientific scene, both in the development of strategies for literature review and in data collection, analysis and triangulation (Souza et al. 2011a, b, 2015).

Souza et al. (2010) underscore that, although researchers use software tools for the analysis of qualitative data, it is essential to take into account that these tools should not "[...] ignore the theoretical, technical and methodological developments that took place in human and social sciences in the last decades." (p. 294).

Based on the epistemological, methodological, technical and technological breakthroughs, we present the combination of two software tools, which can make a difference depending on whether the research is qualitative or qualitative-quantitative. In this study, the first software used is webQDA, which is an online system that supports the analysis of non-numerical and non-structured data. It promoted the identification of categories and/or themes treated by the second software—CHIC— used to support the analysis of similarity across data.

For this study, we sought software that would allow for the systematization and viewing of connections across the different moments in the research. Hence, we chose to use webQDA and CHIC for the treatment of multidimensional statistic data.

3.1 webQDA Software

The webQDA is an online software tool designed to support qualitative investigations mainly for the organization and treatment of collected data (Sousa et al. Souza et al. 2010, 2011a).

These authors consider that, although webQDA is presented as an "empty" form— in the sense that it can be configured to suit the researcher's needs without pointing to a specific design for the investigation—its organizational structure is based on the foundations of content analysis, more specifically in the content structure presented by Bardin (2004): Analysis organization; Coding; Categorization; Inference.

Based on Bardin's (2004) structure, it is important to understand the elements that organize the rationale behind the webQDA work to guide data organization in this study considering three parts: Source, Coding and Questioning.

The insertion of data sources is the first action performed by webQDA, when the researcher inserts the available data (e.g. text, image, video and/or audio). This area can be organized according to the researcher's needs, document type or their function (Souza et al. 2011a, b, 2015). In this study, the sources used and organized to produce the material to be analyzed were: field logs, discussion forum, and written reports created by the teachers attending the development course.

From the textual records in the collected material in the pedagogical interventions (field data collection), it was possible to start the organization of this material on the webQDA.

After this initial organization, we went on to work with the Coding. This step requires a more careful reading of data excerpts aiming to create the themes, the dimensions, the indexes or categories—both the descriptive (a priori) and the interpretive (emerging) ones (Souza et al. 2011a, b, 2015). For this study, we chose to create themes and sub-themes (a priori and emerging from the textual excerpts), which had been organized and coded. These codes consist of numbers (corresponding to the set of themes, dimensions, categories or indexes identified in each phase of the research) and letters, which refer to the mnemonic of the corresponding theme.

Table 1 Themes and sub-themes of the contextual evaluation plan

Code	PCEP—Preparation of CEP by the teacher
	A priori and emerging sub-themes
01PCEP	Students characterization
02PCEP	Curriculum contextualization
03PCEP	Identification of curricular convergence zones
04PCEP	Definition of learning objectives
05PCEP	Identification of competencies and/or skills
06PCEP	Proceedings and criteria for choosing the DLM (emerging)
07PCEP	Activity proposed using the DLM
08PCEP	Support resources
09PCEP	Physical context organization
10PCEP	Digital Learning Material used (emerging)
11PCEP	Evaluation devices
12PCEP	Familiarization with the DLM (emerging)
Code	**DCEP—Development in action over CEP by the teacher**
	Emerging sub-themes
13DCEP	Conceptualization and introduction of the issue in the pedagogical activity
14DCEP	Distribution of pedagogical activities using DLM and support resources
15DCEP	Time and space management
16DCEP	Pedagogical activities applied
17DCEP	Evaluation of results in context after DLM use
18DCEP	Students' attitude (inside and outside classroom)

The treated themes are related to the Contextual Evaluation Plan (Ramos et al. 2004): Preparation of the Contextual Evaluation Plan by the teacher (PCEP) (emerging sub-themes from the textual excerpts and a priori identified from the theory); Development of action based on the Contextual Evaluation Plan by the teacher (DCEP) (emerging sub-themes), as shown in Table 1.

In Table 1, we have 2 themes and 18 sub-themes, which includes a total of 9 sub-themes a priori and 9 emerging sub-themes.

After the establishment of the codes and the creation of themes and sub-themes from the textual excerpts, we started to use the webQDA section entitled Questioning, which provides a set of tools to help the researcher challenge the data and form relation matrices based on the configuration assigned in the previous stages.

In this study, we entered one questioning that corresponds to the combinations/relations of the themes related to the Development of the Contextual Evaluation Plan by the teachers (Table 2).

Once this cross is made, the webQDA automatically launches 0 or 1 values in the matrix cells, thus showing the presence or not (true or false; yes or no) of a priori and emerging dimensions corresponding to the textual records (provided by the research subjects.) In the end, it is possible to export this matrix to a spreadsheet

Table 2 Matrix construction

CEP development		
Questioning: What are the relations established by the teachers in the CEP Preparation and Development?		
Matrix	Theme	Similarity tree
1	PCEP (a priori sub-themes) + DCEP (emerging sub-themes)	1st graph

in .CSV format which is the kind of format of text files used to import/export data from spreadsheets between different programs.

This organization performed by the webQDA software corroborates the methodology proposed by Almeida (2008) and adopted by other researchers (Gras and Almouloud 2002; Souza et al. 2011a, b, 2015; Valente and Almeida 2015) to represent this relation using Excel spreadsheets, which bring the codes for data from the textual records in the first column and, in the first row, the codes for the a priori and emerging sub-themes (e.g. 01PCEP, 02PCEP—see Table 1) and the values zero (0) and one (1) in the cells. After exporting the spreadsheet it is possible to open it in Excel, download it and process it using CHIC.

All this process helps and supports the researcher to identify the analyzed elements as well as the triangulation relations established with the various parts of the investigative project.

However, Souza et al. (2011a, b, 2015) remind us that the webQDA and other qualitative analysis tools alone are unable to make the artificial intelligence processes "find," "interpret" or "discover" result patterns. A critical, creative and analytical researcher will always be needed to interpret data.

Finally, having this structural and organizational overview of the software, and hence corroborating Souza et al. (2011a, b, 2015) ideas, the webQDA allowed us to: work on, organize, distribute and systematize data; establish themes and sub-themes that were treated by CHIC; underscore important aspects stemming from the data and, lastly, help interpret the results.

3.2   CHIC

The software tool named Correspondence and Hierarchical Cluster (CHIC) is based on the foundations of a multidimensional statistic method and implicative statistical analysis (ASI). It is used in qualitative and qualitative-quantitative studies to extract association rules of a set of data (subjects and variables), by providing an association index and a representation of its structuring. Régis Gras started to build it in 1985, as an improvement of the theses developed by Almouloud (1992) and Ratsimba-Rajohn (1992) (Gras and Régnier 2015, p. 23).

The multidimensional analysis can be considered an important tool for research in human sciences, whose qualitative expressions risk being limited to vague

phrases, such as: "teachers said that ..., teachers believe that..., we think that ..."
(Almeida 2008, p. 330).

To avoid this limitation, CHIC enables the construction of graphs and the
viewing of meanings in the interrelated data by using approximations, similarities
and contradictions. Thus, it reveals the conceptions of subjects and provides
information, which is not always available in classic symmetric models (Gras 1996;
Almeida 2008).

CHIC provides four types of data representation. We chose the hierarchic
analysis of similarities because this resource enables the viewing of similarities
and/or dissimilarities between themes and sub-themes of variables organized in
levels through a hierarchic similarity tree (Almouloud 1992).

Aiming to find new possibilities of interpretation for the reflections recorded by
the teachers participating in the PROUCA development course and more specifically in the Development of the Contextual Evaluation Plan, as shown in Table 2,
we adopted the articulation of the matrix organized through the webQDA to generate the similarity tree constructed through CHIC.

Hence, regarding the development of the Contextual Evaluation Plan, we made
one questioning (Table 2)—information presented by the teachers in the CEP
Preparation and Development—which generated the similarity tree graph (Fig. 2).

For the development and analysis of the graph shown in Fig. 2, U-shaped
clusters were identified in the similarity trees, namely classes and sub-classes, and
their interconnections. According to Borges (2009), usually the analysis of similarities starts at the strongest class, that is, with the cluster showing the highest
grade of similarity and represented by the shortest distance in the U-shape width.

Gras and Almouloud (2002) and Borges (2009) explain that it is up to the
researcher to interpret the relations and interconnections viewed in the trees using
the contextual knowledge about the research scope and also by the theoretical basis
of the study, while remaining open to identify unexpected relations leading to
surprising results and new findings about the case in study.

Next, we present the weaving of the result analysis of a section from one
research conducted by Almeida (2013) and Godoi (2013), using the
meta-ethnographic approach.

4   The Weaving of the Meta-Ethnographic Analysis

From the section selected for this study and for the result analysis using the
meta-ethnographic approach, we sought to synthesize the findings of the similarity
tree (CEP Preparation and Development) and integrate them into the analysis.

We consider as findings the strongest U-shaped clusters referred to as classes and
sub-classes, that is, the cluster with the higher grade of similarity identified by the
shortest distance in the U width as shown by the similarity tree in Fig. 1. From
these clusters in the similarity tree, we defined one theme to discuss in this
meta-ethnography: Digital Literacy.

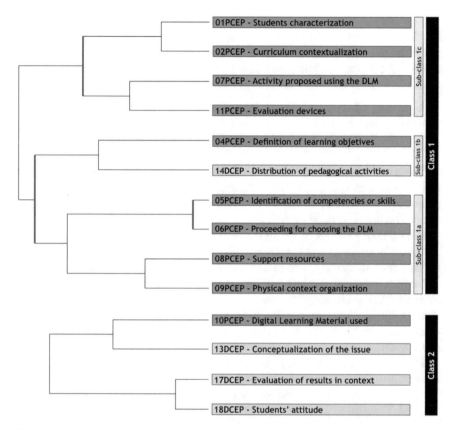

Fig. 2 CEP preparation and development—similarity tree—1st graph

The theme—Digital Literacy—is formed by four sub-themes: Identification of competencies and/or skills [05PCEP], Proceedings and criteria for the choice of the digital learning material [06PCEP], Resources and support material [08PCEP] and Physical context organization [09PCEP], as shown in Fig. 3, which then divides into two sets.

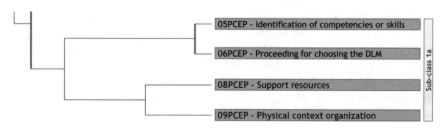

Fig. 3 Digital literacy

The first set of this sub-class indicated the most significant level in the graph (level 1—similarity 0.915095) showing that when the teachers performed the proceedings (actions) and/or when they established pedagogical criteria to choose the digital teaching material, they also tried to identify the competencies that students need to develop. The following excerpt shows this concern:

At the time, I chose to work with the Blog tool and create together with the students this feature to take ownership of this tool, and thus contribute to their digital literacy and enhancing reading and writing skills. I also tried to help students develop the ability of reasoning and communicating, critical thinking and creativity. [01 EST1]

This excerpt also shows that the teacher is concerned about the students' digital literacy and more specifically about the skills that the students need to develop such as reading and writing critically and consciously so that these skills make sense when working with them in the Blog.

Rojo (2009) explains that with the emergence and continuous expansion of the access to Information and Communication Digital technologies (ICDT), there is a demand for new literacy. The author also explains that the term "literacy" seeks to encompass the social uses and practices of language that involve writing in different contexts. Digital literacy, therefore, encompasses reading and writing using ICDT (Soares 2002; Warschauer 2008; Valentini et al. 2013) to develop the skills of rational and critical thinking, creativity, communication, protagonism, cooperation and meaning making.

This first set (sub-class 1a) unfolds and creates the second, in which we can identify the choice of support resources and the physical context organization that the teacher brings to class. The following textual excerpt shows this organization:

The class will be divided into groups of 4 and 5 persons totaling 6 groups. Each group will be responsible for further development of a type of geometric figure. They will be doing this activity in *Impress* and presenting the slides on the digital board. [02EST2]

In both excerpts, the proposal to use the technology in the pedagogical practice is present in distinct situations. In the first excerpt, the teacher—working at elementary school I—expresses his concerns, chiefly, about his students' digital literacy. The second excerpt suggests that the teacher working at elementary school II, besides having concerns about digital literacy, is also preoccupied about other situations, such as context organization, expansion of a specific issue in the curriculum and the organization of this issue to be presented in the classroom.

Therefore, this set of four sub-themes shows that by performing the proceedings (actions) and/or by establishing/adopting criteria for the choice of the digital teaching material, the teacher also seeks to create conditions/strategies for the development of competencies that the students need to develop, such as reading and writing, together with digital literacy. After this step, the teacher starts to organize the physical context for his class and to choose the support resources that will be used in the presentations of students' assignments.

5 Final Remarks

This article analyzed the potential applications of the articulation between software programs for research in education designed to investigate and analyze qualitative data similarities, namely webQDA and CHIC.

For that, we used the meta-ethnographic approach which allowed for the following: retrieve and define a section in a study conducted within a broader investigation (Almeida 2013; Godoi 2013); reformulate and recreate themes from the strongest clusters of one similarity tree; synthesize and integrate the findings of this study aiming to combine them and identify changes; and, finally, to conceptualize results.

As for the synthesis and findings of this research, what was mostly noticeable in the development of the Contextual Evaluation Plan in the similarity tree was the identification of the teachers' need to establish pedagogical criteria as guidelines for the digital teaching material, as well as the need for them to intervene and organize the context and create the conditions for the development of multiple literacy forms by the students.

As for the conceptualization of the combined use of both software tools, it is important to remember that they are not static and each research requires a suitable way to be conducted depending on the object characteristics, context, data to be analyzed, and the researcher's style.

In this study, the combined use of webQDA and CHIC brought significant contributions for the research development, as described throughout the text. It is possible to infer that the greater gain we obtained in the use of webQDA was to organize the research data, edit the themes and sub-themes and make the questioning flexible, besides enabling us to prepare the data to be used by CHIC to answer the research questions. Having all these possibilities, the use of the two software tools added quality to the study and also saved time as webQDA automates the (re)constructions performed. If we had not used it, we would have had to manually construct all the spreadsheets in Excel and then import them to CHIC. The CHIC software, in turn, allowed for the hierarchical structuring of the set of themes and sub-themes and created a rule-based system, identified the recurrences, exceptions, approximations and distancing, which made it more significant for the study, as opposed to simply compiling the information conveyed by each of these classes represented in the hierarchical tree. Hence, it was possible to make conjectures based on stable representations regarding the meaning of relations depending on the level of hierarchy of the classes.

Finally, as a result, we believe we have contributed to the construction of a referential as to how to use both software tools in combination. Therefore, this use does not imply that these tools should be used separately, but rather, that one software is used to (re)feed the work with the other. Hence, we are certain that other articulations with different software require a movement towards new investigations.

Acknowledgments We thank the National Postdoctoral Program (PNPD/Capes) for funding this study and the Brazilian National Council for Scientific and Technological Development (CNPq) for funding the research productivity of Professor Maria Elizabeth Bianconcini de Almeida.

References

Alarcão, I. (2015). "Dilemas" do jovem investigador: dos "dilemas" aos problemas. In Costa, A. P., Souza, F. N., & Souza, D. N. (Eds.), *Investigação Qualitativa*: inovação, dilemas e desafios (pp. 103–121). Porto: Ludomedia.

Almeida, M. E. B. (2008). Mapeando percepções de docentes no CHIC para análise da prática pedagógica. In Okada, A. (Eds.), *Cartografia cognitiva:* mapas do conhecimento para pesquisa, aprendizagem e formação docente (pp. 325–338). Cuiabá: KCM.

Almeida, M. E. B. (Ed.). (2013). *Relatório Técnico Científico. O currículo da escola do século XXI* – integração das TIC ao currículo: inovação, conhecimento científico e aprendizagem. Pontifícia Universidade Católica de São Paulo. São Paulo: CNPq.

Almouloud, S. (1992). *L' Ordinateur, outil d'aide à l'apprentissage de la démonstration et de traitement de donnés didactiques.* Thèse Doctorat, U.F.R. de Mathematiques, Université de Rennes I, Rennes, France.

Bardin, L. (2004). *Análise de conteúdo.* Lisboa: Edições 70.

Borges, M. A. F. (2009). *Apropriação das tecnologias de informação e comunicação pelos gestores educacionais.* Tese de Doutorado, Programa de Pós-Graduação em Educação: Currículo, Pontifícia Universidade Católica de São Paulo, São Paulo.

Costa, A. P., Linhares, R., & Souza, F. N. (2012). Possibilidades de análise qualitativa no webQDA e colaboração entre pesquisadores em educação em comunicação. *3º Simpósio de Educação e Comunicação* (pp. 276–286). Aracaju: Universidade Tiradentes. Retrieved February 1, 2016, from https://www.webqda.com/wp-content/uploads/2012/12/artigo3SimposioEducacaoComunicacao2012.pdf

Figueiredo, A. D., & Afonso, A. P. (2006). Context and learning: A philosophical framework. In A. D. Figueiredo & A. P. Afonso (Eds.), *Managing learning in virtual settings: The role of context.* Hershey, PA, USA: Information Science Publishing.

Godoi, K. (2013). *Avaliação de material didático digital na formação continuada de professores do ensino fundamental:* uma pesquisa baseada em design. Tese de Doutorado, Programa de Pós-Graduação em Educação: Currículo, Pontifícia Universidade Católica de São Paulo, São Paulo.

Godoi, K., & Padovani, S. (2009). Avaliação de material didático digital centrada no usuário: uma investigação de instrumentos passíveis de utilização por professores. *Production Journal, 19* (3), 445–457.

Gomes, M. J. (2004). *Educação a distância:* um estudo de caso sobre formação contínua de professores via internet. Braga: Centro de Investigação em Educação. Instituto de Educação e Psicologia. Universidade do Minho.

Gras, R. (1996). *Nouvelle méthode exploratorie de données.* França: La Pensée Savage, Editions.

Gras, R., & Almouloud, S. (2002). A implicação estatística usada como ferramenta em um exemplo de análise de dados multidimensionais. *Educação Matemática Pesquisa, 4*(2), 75–88.

Gras, R., & Régnier, J. C. (2015). Origem e desenvolvimento da Análise Estatística Implicativa (A.S.I.). In Valente, J. A. & Almeida, M. E. B. (Eds.), *Uso do CHIC na formação de educadores:* à guisa de apresentação dos fundamentos e das pesquisas em foco (pp. 46–54). Rio de Janeiro: Letra Capital.

Lee, R. P., Hart, R. I., Watson, R. M., & Rapley, T. (2015). Qualitative synthesis in practice: Some pragmatics of meta-ethnography. *Qualitative Research, 15*(3), 334–350.

Noblit, G. W., & Hare, R. D. (1988). *Meta-ethnography: Synthesizing qualitative studies.* Newbury Park, CA: Sage.

Nokelainen, P. (2006). An empirical assessment of pedagogical usability criteria for digital learning material with elementary school students. *Educational Technology & Society, 9*(2), 178–197.

Ramos, J. L., Teodoro, V. D., Maio, V. M., Carvalho, J. M., & Ferreira, F. M. (2004). Sistema de avaliação, certificação e apoio à utilização de software para a educação e formação. *Cadernos SACAUSEF*, (1), 21–44. Retrieved February 1, 2016, from http://www.crie.min-edu.pt/files/@crie/1186584566_Cadernos_SACAUSEF_22_45.pdf

Ratsimba-Rajohn, H. (1992). *Contribution à l'etude de la hiérarchie implicative, application à l'analyse de la gestion diddactique des phénomènes d'ostension et de contradiction*. Thèse Doctorat, U.F.R. de Mathematiques, Université de Rennes I, Rennes, France.

Reeves, T. C. (2000). Socially responsible educational technology research. *Educational Technology, 40*(6), 19–28.

Rojo, R. (2009). *Letramentos múltiplos, escolar e inclusão social*. São Paulo: Parábola Editora.

Shapley, K. S., Sheehan, D., Maloney, C., & Caranikas-Walker, F. (2010). Evaluating the implementation fidelity of technology immersion and its relationship with student achievement. *Journal of Technology, Learning, and Assessment, 9*(4). Retrieved April 30, 2016, from https://ejournals.bc.edu/ojs/index.php/jtla/article/view/1610/1460

Soares, M. (2002). Novas práticas de leitura e escrita: letramento na cibercultura. *Educação & Sociedade, 23*(81), 143–160. Retrieved February 1, 2016, from http://www.scielo.br/pdf/es/v23n81/13935.pdf

Souza, F. N., Costa, A. P., & Moreira, A. (2010). WebQDA – Software de apoio à análise qualitativa. *5ª Conferência Ibérica de Sistemas e Tecnologias de Informação*. Santiago de Compostela: Associação Ibérica de Sistemas e Tecnologias de Informação. Retrieved February 1, 2016, from http://www.webqda.com/wp-content/uploads/2012/06/CISTI2010_WebQDADevelopment.pdf

Souza, F. N., Costa, A. P., & Moreira, A. (2011a). Questionamento no processo de análise de dados qualitativos com apoio do software webQDA. *EduSer - Revista de Educação, 3*(1), 19–30.

Souza, F. N., Costa, A P., & Moreira, A. (2011b). Análise de dados qualitativos suportada pelo software webQDA. *7ª Conferência Internacional de TIC na Educação*. Braga. Retrieved February 1, 2016, from http://www.webqda.com/wp-content/uploads/2012/06/artigoChallanges2011.pdf

Souza, F. N., Costa, A. P., & Moreira, A. (2015). Questioning in the qualitative research process. How ICT can support this process?. *4o Congresso Ibero-Americano de Investigação Qualittiva*. Aracaju. Retrieved April 4, 2016, from http://proceedings.ciaiq.org/index.php/ciaiq2015/article/view/166/329

Squires, D., & Preece, J. (1996). Usability and learning: Evaluating the potential of educational software. *Computer and Education, 27*(1), 15–22.

Squires, D., & Preece, J. (1999). Predicting quality in educational software: Evaluating for learning, usability and synergy between them. *Interacting with Computers, 11*(5), 467–483.

Suhr, K. A., Hernandez, D. A., Grimes, D., & Warschauer, M. (2010). Laptops and fourth-grade literacy: Assisting the jump over the fourth-grade slump. *Journal of Technology, Learning, and Assessment, 9*(5). Retrieved April 30, 2016, from http://ejournals.bc.edu/ojs/index.php/jtla/article/view/1610/1459

Valente, J. A. (2015). O uso do CHIC na pesquisa. In Valente, J. A. & Almeida, M. E. B. (Eds.), *Uso do CHIC na formação de educadores:* à guisa de apresentação dos fundamentos e das pesquisas em foco (pp. 79–115). Rio de Janeiro: Letra Capital.

Valente, J. A., & Almeida, M. E. B. (Eds.) (2015). *Uso do CHIC na formação de educadores:* à guisa de apresentação dos fundamentos e das pesquisas em foco. Rio de Janeiro: Letra Capital.

Valentini, C. B., Pescador, C. M., & Soares, E. M. S. (2013). O laptop educacional na escola pública: letramento digital e possibilidades de transformação das práticas pedagógicas. *Educação, 38*(1), 151–164.

Warschauer, M. (2008). Laptops and literacy: A multi-site case study. *Pedagogies., 3*(1), 52–67.

Use of the Software IRAMUTEQ in Qualitative Research: An Experience Report

Maria Marta Nolasco Chaves, Ana Paula Rodrigues dos Santos,
Neusa Pereira dos Santosa and Liliana Müller Larocca

Abstract This article describes the experience of the use of the software IRAMUTEQ to support qualitative research, and thus concepts and some characteristics of the software are addressed to aid in understanding and to illustrate its features. The observations come from two surveys carried out in the city of Curitiba, which were centered on the vulnerability of and the promotion of a healthy lifestyle to adolescent students. The tool enabled the organization of data collected from the participants' speeches acquired through semi-structured interviews. Both studies went through the same steps in the processing of the data from the interviews because, initially, they were transcribed for the preparation of the *corpus* of each of the studies so as to be inserted in the software. Categorization of the data was performed using Descending Hierarchical Classification (DHC). This tool allowed us to show and confirm the thematic categories in the same way that it added quality to the presentation and supported the results.

Keywords Software tools · IRAMUTEQ · Qualitative research · Nursing · Collective health

M.M.N. Chaves (✉) · A.P.R. dos Santos · N.P. dos Santosa · L.M. Larocca
Departamento de Enfermagem, Universidade Federal Do Paraná,
Curitiba, PR, Brazil
e-mail: mnolasco@terra.com.br

A.P.R. dos Santos
e-mail: ana25rodrigues@gmail.com

N.P. dos Santosa
e-mail: neusapesantos@gmail.com

L.M. Larocca
e-mail: larocca_m@terra.com.br

© Springer International Publishing Switzerland 2017   39
A.P. Costa et al. (eds.), *Computer Supported Qualitative Research*,
Studies in Systems, Decision and Control 71,
DOI 10.1007/978-3-319-43271-7_4

1 Introduction

This article reflects on the use of the software IRAMUTEQ in the organization of data and as support in the analysis of said data in two qualitative exploratory searches: "Vulnerabilidade na adolescência: a perspectiva de gestores e líderes do movimento social organizado em um território de Curitiba-PR" ("Vulnerability in adolescence: the perspective of managers and leaders of the social movement organized in a territory of Curitiba, PR") and "Promoção da Saúde do escolar adolescente segundo as diretrizes do programa de saúde do escolar: uma experiência em um município do Sul do Brasil." ("Health promotion to adolescent students according to the school health program's guidelines: an experience in a Southern Brazilian municipality"). The studies were grounded in the theoretical and methodological framework of the Theory of Praxis Intervention in Nursing in General Health and critical epidemiology, both of which are founded in historical and dialectical materialism and thus advise that the praxis in nursing and public health should occur via the dynamic systematization of capturing, interpreting, and intervening in health phenomena, relating this to the processes of production and reproduction of society in its historicity and dynamicity (Egry 1996; Breilh 2006).

To uncover these processes, which are not directly observable in reality, researchers have sought, through the guidance of critical epidemiology (Breilh 2006), to identify the processes of wear and protection present in the lives of the population segment defined for the studies—adolescents living in an area covered by a local health service. To further the discussion and meet the objectives proposed in each survey, data regarding the processes that are directly or indirectly encountered in the lives of those subjects was examined. These processes were related to current public policies that guided actions which are the responsibility of local services in the areas of health, education, sports and leisure, security, and social action. Information about the processes related to the social movement organized by its demands that seeks to establish the necessary confrontations in order to change the local reality was also gathered.

To handle this stage of the studies, in which the particular and structural dimension related to the determination of the local reality was examined, data available in secondary banks on public websites was used. To collect primary data, which was done through semi-structured interviews, it was sought to explore the singular dimension of the reality and objects defined for the referred studies, therefore there were 88 interviews, with 73 participants in the study of health promotion and 15 participants in the study about vulnerability.

Among the participants, managers and professionals of the local services were encountered. These were responsible for developing actions that were, in some way, directed at the adolescents of that region. These health education actions, which were developed in six workshops at the local health care facility, met one of the objectives of the study of health promotion. Adolescent students who partici-pated in these actions were also interviewed. In this phase of data collection, ethical aspects of research with human beings were respected, according to the current resolution in Brazil (Brasil 2012).

Thus, faced with scores of information and the need to discuss in depth the data collected, a qualitative research support software that helps organize and systematize data was used, guided by discourse analysis (Bardin 2011). It is believed that this process enhances the construction of knowledge in studies with little subjective interference from the researchers involved in the presentation of the results, and consequently helps the development of data analysis, since the support of the thematic categorization of the data can be both proven and disproven through textual analysis, which is done by clustering the fragments according to the frequency of the related term or terms, seeing as the system itself performs the data crossing to the extent that the researcher directs the command provided by the software developed to this end.

With the use of the software in the aforementioned studies, the authors sought to overcome some of the challenges identified around the methods and techniques used in qualitative research. At present, the relevance of the development of such research—with emphasis on scientific rigor, transparency, and the inclusiveness of the stages of collection and processing of data so as to ensure the reliability of their analysis, the magnification of subjects/source, etc., and, consequently, the dissemination of knowledge built of the study of means of impact—is debated (Latimer 2005; Egry and Fonseca 2015; Tong et al. 2007).

Such overcoming is in the historical difficulty of acceptance of qualitative studies in contrast to quantitative studies, that, for the most part, according to the chosen methods, can be replicated and permit comparison of results (Sánchez 2015); however, even with greater acceptance, the knowledge constructed from this thinking has its limits and does not allow for certain seizures of objects and realities chosen for the study, especially those that are of interest to the social sciences. Therefore, regardless of the nature of the study, quantitative or qualitative, it is necessary to reflect on these limits and outline studies that help to overcome or complement the acquired knowledge.

2 The Process Organizing of Empirical Data Using the Software IRAMUTEQ in the Discussion and Analysis of Data

Qualitative research explores complex phenomena, which are obtained from non-quantitative data, usually from interviews and focus groups. The interviews—chosen method for the collection of primary data in the aforementioned studies—allowed for the analysis of the experiences and meanings attributed to a particular phenomenon, namely vulnerability of adolescents and health promotion to adolescent students (Tong et al. 2007). And, in order to analyze all of the verbal text material produced, the use of specific software has been increasingly present, especially in studies in which the *corpus* to be analyzed is large (Camargo and Justo 2013a).

In this sense, in order to analyze the amount of *corpus* originating from the interviews, the software IRAMUTEQ (*Interface de R pour les Analyses Multimensionnelles de textes et de Questionnaires*) was used. IRAMUTEQ is a free software anchored in the R software, which enables different processing and statistical analysis of texts—in this case produced from interviews—documents, and other modalities, as it organizes the distribution of vocabulary in an understandable and visually clear manner, thus facilitating the process of organizing the collected material (Camargo and Justo 2013a).

IRAMUTEQ was developed by the French researcher Pierre Ratinaud in 2009, and, despite originally being in French, this software already has full dictionaries in English and Italian, and more recently in Portuguese, since it has been used in Brazil since 2013, and has since become innovated qualitative health research (Lowen et al. 2015; Camargo and Justo 2013b).

This software makes different types of analysis of textual data possible, from the simplest—such as basic lexicography, which covers, primarily, the lemmatization and the calculation of word frequency—to multivariable analysis, such as Descending Hierarchical Classification (DHC), post-factorial correspondence analysis, similitude analysis, which enables a lexical analysis of textual material, providing contexts (lexical classes) characterized by a specific vocabulary e by segments of text that share this vocabulary (Camargo and Justo 2013a). One of the possibilities of this lexical analysis is verified in the word cloud that can be made through the researcher's commands, as shown in Fig. 1. In this example, the relationship between lexical terms and the term adolescent in the study that analyzed vulnerabilities is shown.

The software also "classifies segments of text according to their respective vocabularies and their conjunction is distributed based on the frequency of reduced forms, from matrices, crossing segments of texts and words" (Camargo 2005). In this phase of organization of empirical material, the software aims to obtain Elementary Context Unit (ECU) classes from the Initial Context Units (ICU) as the program performs the scaling of segments of text or ECU, which have an average of three lines, classified according to the most frequent vocabulary and the highest chi-squared values in the class, considering the understanding that they were significant for the qualitative analysis of the data (Camargo and Justo 2013a; Ratinaud 2009).

In the development of the studies, the *corpus* was prepared, which is the set of Initial Context Units (ICU) to be analyzed, and in this sense, the development of the *corpus* resulted from the transcription of the interviews and, later, was grouped into a single text file using the software OpenOffice.org and saved as text.txt, separated by command lines according to the research variables and, from the *corpus* prepared for the study proceeded the textual analysis, being that, in both studies, the method used was the Descending Hierarchical Classification (DHC), as the *corpus* of each study worked with a textual set centered on a theme (Camargo 2005).

The ICUs or texts were build based on each guiding question of their respective studies and not on each interview with the participants, i.e., care was taken with the

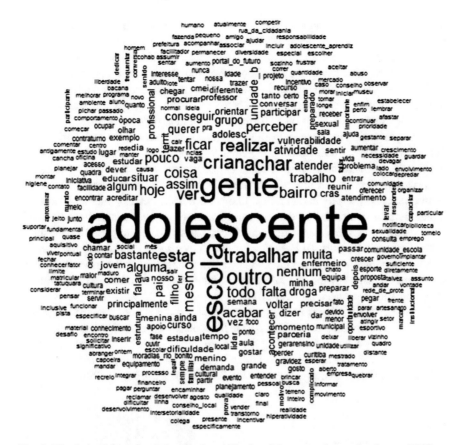

Fig. 1 Word cloud from the study of vulnerability in adolescence. *Source* dos Santos (2015)

preparation of the *corpus* through questions asked to each participant to not use the complete interviews. This choice was made to facilitate the organization of the empirical material and, later, to facilitate the process of analysis of the results. Therefore, the participants' speeches were grouped according to the category to which they belonged (manager, employee, teacher, adolescent) and, thus, the issuer was evidenced by means of encoding previously established by the researcher (Camargo and Justo 2013a; Camargo 2005).

From the analysis performed by the software, which is done by commands from the researcher, IRAMUTEQ organized the classes, which are composed of several segments of texts according to a classification based on the distribution of vocabulary, originating from a dendrogram of the DHC from the *corpus* that illustrated the relationship between these. IRAMUTEQ allowed for the description of each of these classes, chiefly through their characteristic vocabulary (lexicon) and through the words with asterisks (variables).

The command lines and the variables containing asterisks were: **** *n_01, in an increasing ordinal sequence that would meet the category and the number of respondents; *gest_1, manager; *educ_1, educator; *saud_1, health; *soc_1, social action; *adolesc_1, adolescent student; *func_1, school employee from the area studied; *prof_1, high-school teacher from the area studied. It should be noted that this procedure, identifying the dendrogram of the DHC, lasted 12 s for the study that worked with vulnerabilities in adolescents and 25 s for the study of health promotion, which was a significant advantage at this stage of the studies when comparing the procedure to other forms of qualitative data processing.

After processing the *corpus*, from which the dendrogram of the DHC originated, the software showed the results in another way, or in other words, presented the results using a factorial correspondence analysis made from the DHC, seeing that the program, through calculation, provided the most characteristic segments of text from each class (*corpus* in color, also known as *corpus* cooler), which allowed the researchers to contextualize the typical vocabulary in each class. Based on this contextualization, the classes originated in each study were interpreted by means of content analysis (Bradin 2011), which is defined by a set of communication analysis techniques, which use systematic and objective procedures to describe the content of the messages, allowing the inference of knowledge related to the condition of receiving such messages (Bradin 2011).

In the study of vulnerability in adolescence, 74 texts were analyzed and processed by the software, from which 510 text segments were obtained; of these, 455, or 89.22 %, were used. After sizing the text segments, classified according to their vocabularies, the text classes were defined, as shown in Fig. 2, in which the dendrogram of the Descending Hierarchical Classification (DHC), is illustrated.

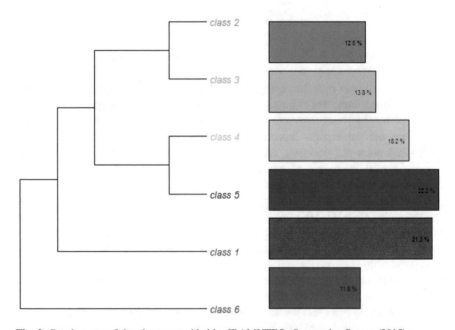

Fig. 2 Dendrogram of the classes provided by IRAMUTEQ. *Source* dos Santos (2015)

According to Fig. 2, the *corpus* was divided into four *sub-corpora*. Class 6, which consists of 54 Elementary Context Unit (ECU) and which concentrates 11.9 % of the total ECUs of the *corpus*, was obtained from one of the *sub-corpora*. Class 1 was obtained from another *sub-corpus*, with 97 ECU, which corresponds to 21.3 % of the ECUs, and, from this, three more divisions from which classes 2, 3, 4, and 5 originated, with 57, 63, 101, and 83 ECUs and corresponding to 12.53, 13.8, 22.2, and 18.24 %, respectively, of ECUs from the whole *corpus*. For each class, a list of words generated from chi-square tests (X2) was computed, or rather, this analysis aims to obtain the ECU classes, which present similar vocabulary to each other and, at the same time, different vocabulary from the ECU of the other classes, that is, as explained above, the program performs calculations and provides results that allow for the description of each of the classes, mainly through their characteristic vocabulary (lexicon). The percentage, on the other hand, refers to the occurrence of the word in the text segments in that class in relation to its occurrence in the *corpus*, while the chi-square refers to the association of the word with the class (Camargo and Justo 2013b) (Table. 1).

After reading and analyzing the *corpus*, it was noted what classes 2, 3, 4, and 5 had the same logical sequence of subjects demonstrated by the software, that is, after processing the *corpus* and reading thoroughly the material, by means of the *corpus* cooler, analysis of the results was done through content analysis (Bradin 2011). From this analysis, classes 2 and three were joined together, as well as

Table 1 Distribution of the terms by chi-square (X2) and the frequency of the term in the classes

Class	Word	X2	%
Class 1	You	63.34	65.31
	Phase	41.6	100.0
	Prepared	37.74	90.91
Class 2	Community Health Center	86.91	80.0
	Nurse	63.2	90.91
	Social Assistance Regional Council (CRAS)	37.79	62.5
Class 3	School Community	44.24	100.0
	Accomplish	41.76	44.68
	Sports Field	37.03	87.5
Class 4	Family	41.33	73.08
	Also	33.76	52.73
	People	32.11	43.0
Class 5	Neighborhood	57.6	62.5
	Grow	31.86	100.0
	Demand	30.55	81.82
Class 6	Violence	138.22	95.0
	Drug	115.9	82.61
	Pregnancy	99.38	100.0

Source dos Santos (2015)

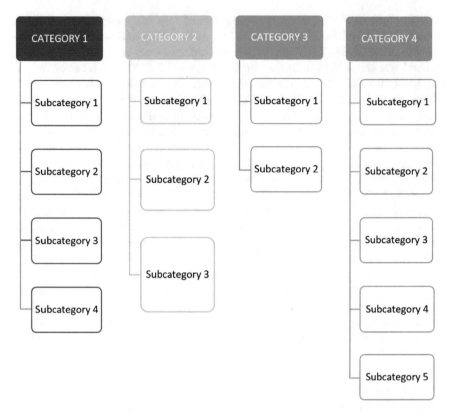

Fig. 3 Diagram of thematic categories and subcategories demonstrated in the participants' speeches in a study. *Source* dos Santos (2015)

classes 3 and 4, thus forming four major thematic categories, which were based on the theoretical framework that guided the study, as shown in Fig. 3.

Content analysis (Bradin 2011) consists of phases that will help the researcher organize the material, and here, it is worth mentioning that the organization of said material was done using the formatting of the *corpus* to be analyzed by the software IRAMUTEQ, and, with the disclosure of the classes, we proceeded to the stage of superficial reading and defining of analysis categories, which was done through analytical categories defined for each of the studies. The objective of this phase was to go beyond common sense and subjectivity in the interpretation of the data collected and, thus, interpret critically the participants' speeches through the identification of core meanings in each thematic category shown. The core meanings were identified in each class originated in the dendrogram of the DHC, as well as the relationships between the classes and the frequency of texts in each of these.

Through the experience of the studies, the core meanings evident were: violence, drugs, pregnancy, STIs, vulnerability, gender and generation, territory, and intersectionality, whose presence and frequency had greater chi-square values and significance $p > 0.001$, indicating a significant association for the proposed objectives. The chi-square test is used to verify the association of ECU with a particular class; therefore, the higher the value, the greater the association, whereas all of the selected words had a $p > 0.001$, indicating a significant association (Chartier and Meunier 2011).

It is believed that statistical calculations of qualitative variables, from the entered texts, do not make the software a proper research method, since this does not analyze the organized data; however, in the process of organizing data, this tool allows for more thorough exploration of data. Therefore, the use of software in qualitative studies can facilitate the processing of long and numerous texts, but it does not replace the essential role of the researcher in the collection, preparation, and analysis of data. The outline of the study, the methodological approach, the interpretation, and the analysis will always be the responsibility of the author of the study, who should respect the ethical aspects and quality parameters of qualitative research (Egry and Fonseca 2015; Camargo and Justo 2013a; Chartier and Meunier 2011; Lahlou 2012; Coutinho 2015).

3 Conclusion

The use of the software IRAMUTEQ was decisive for the processing of the research data, as it allowed for the development of a more critical look at the material—through the frequency of the terms, and the possibility to confirm the participants' speeches using the core meanings in each thematic category outlined by the DHC dendrogram. This process allowed for the qualification, the consistency, and the visualization of empirical texts in the categorization process of the speeches and, consequently, the qualification of the results and the analysis done in both studies.

Although the software's interface is relatively easy to use, it is important to note that the free access of this tool has contributed significantly to its use in research. Another relevant issue to be emphasize is the role of the researcher as of fundamental importance in the design of the study, organization of collected material, and the analysis process, as the software does not do the research itself, but aids in the organization, processing, and in the support of the findings to carry out data analysis, mainly in studies with large volumes of text and, for this to happen, it is necessary that the researcher has knowledge for the use of the selected tool.

References

Bardin, L. (2011). Análise de conteúdo. São Paulo: Editions 70.
Brasil. (2012). Resolução 466/2012. Regulatory guidelines and standards for research involving humans. Ministério da Saúde/Conselho Nacional de Saúde (Ministry of Health/National Health Council), Brasília, December 12, 2012.

Breilh, J. (2006). *Epidemiologia crítica: Ciência emancipadora e interculturalidade*. Rio de Janeiro: Hucitec.

Camargo, B. V. (2005). ALCESTE:um programa informático de análise quantitativa de dados textuais. In A. S. P. Moreira, B. V. Camargo, J. C. Jesuíno, & S. M. Nóbrega (Eds.), *Perspectivas teórico-metodológicas em representações sociais* (pp. 511–539). João Pessoa, PB: Editora da Universidade Federal da Paraíba.

Camargo, B. V., & Justo, A. M. (2013a). IRAMUTEQ: um *software e* gratuito para análise de dados textuais. Rev. Temas em psicologia, Ribeirão Preto, Vol. 21, no. 2, pp. 513–518. Available at: http://pepsic.bvsalud.org/pdf/tp/v21n2/v21n2a16.pdf. Accessed on: 10 May 2015.

Camargo, B. V., & Justo, A. M. (2013b). Tutorial para uso do software de análise textual IRAMUTEQ. Universidade Federal de Santa Catarina. Available at: http://www.iramuteq.org/documentation/fichiers/tutoriel-en-portugais. Accessed on 10 May 2015.

Chartier, J. F., & Meunier, J. G. (2011) *Text mining methods for social representation analyses in large corpora*. Papers on Social Representations, Paris (FR), Vol. 20, no. 37, pp. 1–47. Available at: http://www.psych.Ise.ac.uk/psr/

Coutinho, C. (2015). Avaliação da qualidade da investigação qualitativa: algumas considerações teóricas e recomendações práticas. In de Souza, F. N., de Souza, D. N., & Costa, A. P. (Eds.), Investigação qualitativa: inovação. Dilemas e desafios. Aracaju: EDUNIT.

Egry, E. Y. (1996). *Saúde coletiva: construindo um novo método em enfermagem*. São Paulo: Editora Ícone.

Egry, E. Y., & Fonseca, R. M. G. S. (2015). Acerca da qualidade nas investigações qualitativas em enfermagem. In de Souza, F. N., de Souza, D. N., & Costa, A. P. (Eds.), Investigação qualitativa: inovação. Dilemas e desafios. Aracaju: EDUNIT.

Lahlou, S. (2012). *Text mining methods: An answer to Chartier and Munier*. Papers in Social Representations, London (UK), Vol. 20, pp. 38.1–39.7. Available at: http://psych.Ise.ac.uk/psr/PSR2011_39.pdf

Latimer, J. (2005). *Investigação qualitativa avançada para a enfermagem*. Lisbon: Sersílito.

Lowen, I. M. V., Peres, A. M., Crozeta, K., Bernardino, E., & Beck, C. L. C. (2015). Competências gerenciais dos enfermeiros na ampliação de estratégia saúde da família. RevEscEnferm USP, São Paulo, Vol. 49, no. 6, pp. 967–973. Available at: http://www.scielo.br/pdf/reeusp/v49n6/pt_0080-6234-reeusp-49-06-0967.pdf

Ratinaud, P. (2009). IRAMUTEQ: Interface de R pourLesanlysesmultidimensionnlles de textes et de questionnaires [computer software]. Available at: http://www.iramuteq.org

Sánchez, C. (2015). Orígenes y evolución de la investigación cualitativa en educación. In de Souza, F. N., de Souza, D. N., & Costa, A. P. (Eds.), Investigação qualitativa: inovação. Dilemas e desafios. Aracaju: EDUNIT.

dos Santos, A. P. R. (2015). *Vulnerabilidade na adolescência: a perspectiva de Gestores e Líderes do movimento social organizado em um território de Curitiba-PR Dissertação (mestrado) – Programa de Pós-Graduação em Enfermagem*. Setor de Ciências da Saúde: Universidade Federal do Paraná.

Tong et al. (2007). Consolidated criteria for reporting qualitative research (COREQ): A 32-item checklist for interviews and focus groups, Vol. 19, no. 6, pp. 349–357. Available at: https://intqhc.oxfordjournals.org/content/intqhc/19/6/349.full.pdf

Implementation of Focus Group in Health Research

M.C. Sánchez-Gómez and M.V. Martín-Cilleros

Abstract Focus groups are a tool for the collection of information that allow to understand the participants' perspective about phenomena that surround them from the approach of the qualitative methodology, moving forward in deepening in their experiences, opinions and meanings. In recent years, an increase in the use of this technique has taken place in the field of health sciences. This article presents the main points to be considered in terms of planning, organization and analysis of collected data by means of focus groups, through the presentation of three examples applied in different studies related to health research, in the pursuit of a systematization and objectification which encourage a greater accuracy in the use of this tool in fields related to psychosciences and their field researches.

Keywords Focus groups · Qualitative methodology · Health sciences research · Qualitative approach

1 Introduction

Research is like a path to the development of knowledge. Carrying out an accurate research will be the key for this way to be appropriate and to achieve the aim.

The first step starts with an inquisitiveness which may be caused by several needs: by the existence of gaps in scientific knowledge about the subject, as a diversion of formulated theories which require empirical validation, by the emergence of new or unknown situations hitherto, as a result of previous researches which open up new ways to look into, even from the personal experience of the researcher (Sánchez-Gómez et al. 2013). When it is decided to eliminate this

M.C. Sánchez-Gómez (✉) · M.V. Martín-Cilleros (✉)
Department of Didactics, Organization and Research Methods,
University of Salamanca, Salamanca, Spain
e-mail: mcsago@usal.es

M.V. Martín-Cilleros
e-mail: viquimc@usal.es

inquisitiveness, some criteria of relevance must be applied, clarity, effectiveness, accuracy and viability, in a manner that the problem may become a searchable issue. On top of some other questions, once the problem is clarified, together with those who are going to take part in the resolution, the next step is to decide between different methodological options.

Looking at the methodological strategies, Buendía and Carmona (1984, in Buendía 1992) reduced the existing possibilities to three possibilities: observation methods, survey methods and measuring methods. The first methods are more related to qualitative research of a naturalistic kind and they use mainly qualitative data, while the latter need quantifying and they are typical of experimental methodology.

Qualitative research was used in the various disciplines of the social sciences since time immemorial. However, in recent years there is increasing interest in the use of these methods in the field of public health (Mallik 2014). According to Mira et al. (2004), this is due to the need of broaching achievable aspects with difficulty from quantitative approaches such as establishing the social impact of certain political decisions, identifying necessary changes in professional functions, agreeing on decision-making about active policies, analyzing the doctor-patient relationship or identifying those aspects which most affect the different interest groups.

Following the humanist-philosophical paradigm, qualitative approaches recognize the uniqueness of the subject, the intersubjective nature of empirical research and the solidarity with the ones polled groups, on the basis of fundamental equality of the human being (Minayo and Guerriero 2014). Through qualitative research, where the world of clichés is not as important as it is in quantitative research, researchers try to understand the participants' perspective (individual or as a group) about phenomena that surround them, delving into their experiences, perspectives, opinions and meanings (Hernández et al. 2010). Therefore, it seems logical the use of qualitative research techniques in health sciences.

In the field of health, the focus group technique for research and evaluation is relatively recent; revealing that is growing exponentially since the beginning of the century (García Calvente and Mateo Rodríguez 2000). However, we are still in a phase where most of the knowledge about focus groups comes from personal experience instead of a systematic research (Coule 2013). Thus, this article tries to encourage and to ease the systematicity of researches based on focus groups in the field of health, proposing a general organizational framework as a guide for the research staff when they start with the task of discourse analysis through these group interviews, encouraging informed decision-making about how researchers can be related to the focus group method.

To this end and first of all, relevant aspects are presented in the gathering of information through focus groups, carrying on with the description of the treatment of gathered data and the process of analysis. Finally, the theory is exemplified through its implementation in health research projects, where authors have cooperated, and also it has allowed for perfection and improvement of the systematization of the process.

2 Key Concepts of the Focus Group

The main purpose of this technique is to understand why and how people think or feel the way they do, without aiming to reach any consensus. The similar and the different aspects of the participants' experiences count with the same importance (García Calvente and Mateo Rodríguez 2000). The interlocutor does not only express freely his opinion regarding his perceptions, but also the views are enriched with the group interaction provided by the focus groups. Therefore the information is influenced by the presence of the others; it is the social interaction in particular, in the framework of a specific dynamics, which offers the quality and the value of the gathered information. Thus, it is required to count with a situation of interaction within a group that makes possible the expression or verbalization of views, information, beliefs, positions of influence or leadership, etc. that can be subsequently analyzed by the researcher. This information is never equivalent to the sum of the individual information expressed by each of the group members.

2.1 Profile of the Participants in the Focus Group

As for the sample model of the discussion groups, it should be noted that it is not statistical, but structural, that is to say, depending on the variables of social structure (they are the ones that influence on the discourse) that could be more important for the aim of the study. It is assumed that a person that belongs to a particular group can be substituted, what concerns the discourse, by another person of that group who possesses the same characteristics. The structural sample aims to concentrate on the relations, not on the individuals. Hence, it is not the statistical representation that matters, but the structural representation does.

There does not exist, as occurs in a statistical sample, a formula that can be used to specify the number of persons that have to be interviewed or how many groups have to be created. The analysis, made by the researcher, of the texts that have arisen from the focus groups will point out if the group provides new information or if instead, what is being said has been already expressed in a group of the same features and there is no input of new ideas. In the first case, the creation of a new group can be considered; in the second case, what is called in qualitative research a saturation of speech is reached.

The selection of participants can be carried out on different manners, from employing professionals to take care of the search, to counting on friends, family or colleagues or to make random phone calls. In any case, we have to keep in mind that the selected persons must not know the researcher or the moderator of the discussion group. Nor should they be acquainted with the objectives of the research or join the group with a planned view and a prepared speech to please the researchers.

The profile of the participants should be established. Therefore, each created group will receive a numbering (G1, G2,..., G6), as well as each profile that will

Table 1 Profile of the participants of focus group number 1

Participant	Profile
G1P1	Paediatrician (Primary Health Services)
G1P2	Nurse (Primary Health Services)
G1P3	Psychologist with experience in the field of developmental disorders
G1P4	Professional from Social Services
G1P5	Professional Social Services
G1P6	Developmental disorders expert
G1P7	Researcher in the field of developmental disorders
G1P8	Another professional (Social or Health Services)

determine the group's composition. This can be observed in Table 1, where the composition of the participants of Group 1 of a research study in health sciences is shown.

The code of each of the participants is very important, so each of his interventions should be identified with his code. This allows us to follow the complete discourse of each participant, besides knowing the point of view of the different groups. We should take into account that in all the created groups, each profile will be always identified with the same number (for example the psychologists will always have number 5, the experts will have always number 8 in all the groups, etc.). In addition, other identification variables, as sex, age, work experience, etc. can be used.

The outline of the meeting will be previously worked out by the research group, with the aim that the received information will be useful to reach the research goals. The first question should be open and the introduction that will be given at the start of the session may not determine the answer nor the type or structure that should be used.

On the other hand, we may not forget the ethical standards related to the qualitative research process. Apart from the informed consent, the privacy and confidentiality principle, we have to take into account the ethical dilemma that aggravates and increases in qualitative research because the interaction with persons and settings is probably more intensive. Each time that a group meeting has taken place, Conde (1996) advises to elaborate, immediately after the debate and discourse, a small essay with the perceptions we have of the group.

2.2   Dynamics of the Focus Group

The general criteria that enable a focus group to function and to be used for the research goals are presented:

- Composition and number of groups

- The totality of the groups has to show diversity, the whole of the views concerning the studied phenomenon.
- Each group has to be fairly homogeneous (this fact encourages interaction) and has to manage one of the possible expected discourses on the phenomenon
- The groups have to be segmented by those variables that homogenize and that make the difference in the discourse (age, sex, social class, ethnicity ...)
- For each segmentation variable, 2 groups are created in order to obtain the saturation of each discourse.

- Size of the group

  - The number of members should not be less than 7 or more than 10.
  - Smaller groups would reduce the interaction.
  - Large groups complicate the management of the group as well as the participation of all the members.
  - The possible absence of a participant should be taken into account.

- Duration of the group session

  - It should be known by the members beforehand.
  - One hour and a half is an appropriate duration.

- The setting/place of the meeting

  - Comfortable, quiet and accessible.
  - The seats should be positioned in a way that encourages group interaction.
  - There should be no symbolic elements, separated from those contexts that could determine the discourse. If it is not possible to avoid symbolism of the place, it should interfere as less as possible and the influence of the symbolism should be analyzed.
  - The identification of the participants. Agreed previously: real or figurative.

- Role of the moderator

  - Abilities for the coordination of the group.
  - Maintain in every moment the authority which will allow the dynamism of the group.
  - Intervene in situations that complicate the group work (the group remains silent or gets annoyed, gets strayed from the specific topic, a participant monopolises the meeting).
  - Participate and speak as less as possible, "to be all eyes and ears".
  - He should not judge on a topic
  - He should control very well his nonverbal communication, as it can transmit meanings and it can be as intrusive as verbal language.
  - A good interviewer should be as a "screen" that sees, listens and thinks, but only says the essential and does not participate in the discussion, and does not judge.

- Records and transcription
  - Record the group sessions.
  - In the transcription the participants should be identified.
  - The transcription has to be literal.
  - There should be a review of the group process once finished the session.
  - Write a report with a description of the participants, attitude, participation and interaction between the participants and the impressions of the interviewers.
  - The review of the process and the early analysis of the first group sessions will help to reflect and to introduce modifications in the outline.

The development of the group session is very important. Therefore, it should be planned properly, which requires some time.

2.3 Data Collection and Information Transcription

When data are recorded using audiovisual means, the transcription is a necessary step for their interpretation. Systems or rules for standardized transcription are not yet available, but some recommendations that all researchers take into account are. What seems to be clear is that we should transcribe the amount, and with the accuracy that the research enquiry requires. The transcription must be literal and as accurate as possible. At present, there exist several computer programmes that can make this part of the work of the researcher less tedious and easier; one of them is Nvivo11.

For the data collection and for the information transcription it is very useful to use a template in an Excel file as the one presented in Table 2.

2.4 Data Analysis

The discourse analysis implies a task of reconstruction of the text meaning. This requires the reading of the text, and its interpretation in its "totality". Before any kind of thematic analysis, of contents or categories, as is usual, the reading of the text should cause a first impression which should precede the analysis. To analyze the discourse of the discussion groups, in a first stage, to detect the discursive positions and the symbolic configurations of the texts, we will follow the analysis guidelines developed by the hermeneutical approach according to the model of Conde (1996). In a second phase, we can use the analytical scheme of Miles and Huberman (a) data reduction; (b) data organization and processing; and (c) achievement of results and verification of conclusions).

Data reduction consists of the selection and summary of the information. It can take place beforehand (when the theoretical base is established, the questions defined, the participants and the means for the gathering of information selected), or

Table 2 Template example of data collection for its transcription

Name focus group: (G1)/ (G2)/(G3):	Date of focus group meeting (dd/mm/yyyy):							
	Duration of the meeting:							
Variable name	Partn_ID	Partic_ID	Age	Sex	Prof_Exp	Job	Fam_Rel	Living
Label description	Labels corresponding to each country	Participant code in the focus group	Age in years at the time of participation in the focus group	Sex of individual (0 = M, 1 = F)	Professional experience	Current main job	Relationship to the person with ASD	Is the family living with the person with ASD? Yes/No
Variables for participant identification		G3P1						
		G3P2						
		G3P3						
		...						

Transcription

Date of transcription (dd/mm/yyyy):

Partic_ID	**Speech**: What everyone says during the focus group meeting

when the researcher summarizes or outlines the data once he has gathered them. The data reduction tasks constitute rational procedures that consist of cataloguing and coding; identifying and distinguishing the units of meaning.

Data processing tries to achieve an organized whole of information; normally this information is presented in a spacious, effective and operational structure that enables to solve issues related to the research. When, in addition, the process involves a change in the language used to express the information, we talk about *data transformation*. One of these procedures are graphics or diagrams, they enable data presentation and the analysis of profound relations and structures within the data. Miles and Huberman (1994) consider the matrix structure where different types of information can be demonstrated (texts, quotes, abbreviations or symbolic elements) and distinct formats. For this task we will use the program for qualitative data analysis, CAQDAS Nvivo11. The program cannot replace the abilities of the researcher we have shown before, but it can help in the practical phase of the analysis as: marking and coding of the text, relation between categories and individuals, elaboration of typologies and profiles, or review, search and recovery of codified units. This computer program supports the processes of deductive or inductive categorizing, and even both of them. This ability enables to design a priori a category system developed from theories that already exist or that are created for the aim of the research, or hypothesis or premises that have been included during the gathering of information. One of the features of this program is that we can classify and organize these categories hierarchically. This enables the researcher to study the relations that exist among the numerous views or categories that he has dealt with, and compare them through specific factors (Boolean, contextual, negative, inclusive, exclusive) to outline conclusions related to the research topics.

The *achievement and verification of conclusions* implies the reconnection of the elements that were distinguished in the previous processes, in order to create a structured and significant whole. To assure the quality of this process, the qualitative analysis will take into account several guidelines as credibility, transferability, dependence, provability, educational authenticity and justice or equity. The conclusions deal with the results, the elements of the research and the interpretation of these by the persons involved. The results will progress in the explanation, the comprehension and the knowledge of reality; and they will contribute to the theorization or intervention of it. Color figures will appear in color in the eBook but will be printed in black and white, therefore make sure that the interpretation of graphs does not depend on color.

3 Example of the Process in Health Research Projects

We hereby, in Table 3, present the previously explained process, developed in research with focus groups in the field of Health Sciences

Table 3 Development of the research with focus groups in various projects in the field of health

Subject research project	Labor importance, detection, prevention and intervention of suicidal behavior
Population profile: Structural sample pattern	56 professional participants of the health area who take part in prevention, attention and intervention of suicidal behavior, they are divided into 8 focus groups
Outline: It shows the aspects that they want to know about the research topic	Outline quote – In your opinion, what grade of importance is given to clinical practice in suicide attempts? Are other pathologies given more importance? – What has been done in the prevention of suicidal behavior? – What are the resources that exist and facilitate the intervention process?
Data transcription: For each component of the outline a range of references is obtained	Transcription excerpt: Internal aspects\\primary assistance G1>-119 encoded references on intervention [range 19, 11 %] Reference 17—range 0, 11 % 29 | 17:40, 5/17:45,6 | You are fearful to deal with what you are not trained for, and not for what you are trained for | G1P2
Transformation of the text into data and coding or assignment of the textual space to the corresponding category, of the gathered information. The concept map of the categories is created in function of the goals of the study and of the question protocol used for the guidance of the debate. The computer program NVIVO 11 is used to support the data analysis	Once the group discourses have been recorded, the categorizing will take place. The categorizing can be shown on different ways, as in a concept map
Data analysis: The speech bubble, the concept map, the contents analysis and the text quotes that are created after the encoding of the information represent the ideas taken from each of the focus groups. These lead to research results	In the speech bubble are demonstrated those words that receive the main frequency percentage In the bar diagram, the text is encoded in categories

Subject research Project	Perceptions of mental health professionals, relatives and patients regarding a long lasting antipsychotic taken by injection
Population profile: Structural sample pattern	48 professional participants of the area of psychiatry, patients diagnosed of schizophrenia and relatives of the patients, they are divided into 7 focus groups
Outline: It shows the aspects that they want to know about the research topic	Outline quote: – Do you think that medication taken by injection might be coercive for patients? In which cases? – What are the advantages and disadvantages of an antipsychotic taken by injection, compared to oral antipsychotics? – Have you noticed any kind of effect on the number of relapses and hospitalizations?
Data transcription: For each component of the outline a range of references is obtained	Transcription excerpt: Internal aspects\\nursing G2>-9 encoded references on the Importance of antipsychotic drugs in schizophrenia [range 1, 96 %] Reference 8—range 0, 24 %

	191	23:09, 1 23:14, 0	A relapse can occur frequently, the problem is with those who are not aware of their disease	ENF2 P7

Transformation of the text into data and coding or assignment of the textual space to the corresponding category, of the gathered information. The concept map of the categories is created in function of the goals of the study and of the question protocol used for the guidance of the debate. The computer program NVIVO 11 is used to support the data analysis	Coding bands, distinguished by colors assigned to the nodes
Data analysis: The speech bubble, the concept map, the contents analysis and the text quotes that are created after the encoding of the information represent the ideas taken from each of the focus groups. These lead to research results	Organizational key word chart and cluster analysis to gather resources or categories that share similar words, values of features of the subjects / resources or coding

Subject research project	Recognition of strengths and weaknesses in early detection, diagnosis and intervention of autism in Europe		
Population profile: Structural sample pattern	190 participants from 10 EU countries who belong to distinct public and private areas related to autism, they are divided into 32 focus groups		
Outline: It shows the aspects that they want to know about the research topic	Outline quote:		
	– How do you think are the existing servicesworking on early intervention?		
	– What factors might improve intervention?		
	– Are there institutional variables that are having influence?		
Data transcription: For each component of the outline a range of references is obtained	Transcription excerpt:		
	Internal aspects\\FR_G2>- 13 encoded references on needs [range 6, 19 %]		
	Reference 4—range 0, 22 %		
	45	22:32 – 23:33	Some families have to go to court and sometimes lose in order to have a diagnosis accepted	FR2P5
Transformation of the text into data and coding or assignment of the textual space to the corresponding category, of the gathered information. The concept map of the categories is created in function of the goals of the study and of the question protocol used for the guidance of the debate. The computer program NVIVO 11 is used to support the data analysis	Graphic of resources or categories to recognize the connections among them		
Data analysis: The speech bubble, the concept map, the contents analysis and the text quotes that are created after the encoding of the information represent the ideas taken from each of the focus groups. These lead to research results	Radial and ramified map comparing nodes counting the number of references		

4 Conclusions

The steady increase in recent years of the number of publications in the area of health using the focus group technique requires the development of a model that systematizes the process through different constituent phases of qualitative research.

Throughout this article a systematic process of organization, planning and data collection by means of the focus group is provided. In this process the participants and their socio-demographic features are identified in order to connect them with the expression of their perceptions, beliefs and attitudes towards the subject of study and it also supports the transcription process.

By the development of different research projects, in which the authors have participated and with which the process is exemplified, a logical sequential scheme of the data collection is obtained. This scheme supports the analytical process of categorizing and coding the information on a systematic manner in order to connect the results that are in accordance with the referential context of each case.

Supported by the computer program CAQDAS Nvivo11, conclusions can be drawn from hierarchical clusters and organizations of the categories established a priori based on the research subject, or hypothesis or concepts that have been incorporated during the gathering of information.

References

Buendía, L. (1992). *«Técnicas e instrumentos de recogida de datos,»* de *Investigación educativa* (pp. 201–248). Sevilla: Alfar, S.A.

Coule, T. (2013). «Theories of knowledge and focus groups in organization and management research", Qualitative research in organizations and management,» *An International Journa, 8,* nº 2, pp. 148–162.

García Calvente, M., & Mateo Rodríguez, y I. (2000). «El grupo focal como técnica de investigación cualitativa en salud: diseño y puesta en práctica,» *Atención Primaria, 25,* nº 3, pp. 3–28.

Hernandez, R., Fernández, C., & Baptista, y P. (2010). *Metodología de Investigación McGraw Hill*. Mexico: McGraw Hill.

Mallik, S. (2014). «Application of qualitative research in public health,». *Journal of Comprehensive Health, 2*(2), 73–82.

Milles, M. B., Huberman, A. M., & Saldaña, y J. (2013). *Qualitative data analysis: An expanded sourcebook,* 3rd ed., Arizona: Sage.

Minayo, M. C. d. S., & Guerriero, y. I. C. Z. (2014). «Reflexividade como éthos da pesquisa qualitativa,» *Ciência & Saúde Coletiva, 19,* nº 4, pp. 1103–1112.

Mira, J., Pérez-Jover, V., Lorenzo, S., Aranaz, J., & Vitaller, y J. (2004). «La investigación cualitativa: una alternativa también válida,» *Aten Primaria 2004, 34,* nº 4, pp. 161–169.

Sánchez Gómez, M. C. (2015). «La dicotomía cualitativo-cuantitativo: posibilidades de integración y diseños mixtos,» *Campo Abierto, 1,* nº 1, pp. 11–30.

Sánchez Gómez, M. C., & Martín Cilleros, y M. V. (2015). «Contextualización de la investigación cualitativa: de la confrontación al continuum,» de *Conferencia Inaugural 1º Congresso Educaçao, Pedagogía & Innovaçao*, Castelo Branco.

Sánchez-Gómez, M. C. (2010). *«Técnicas grupales para la recogida de información,»* de *Principios, métodos y técnicas para la investigación educativa* (pp. 223–245). Salamanca: Dykinson.

Sánchez-Gómez, M., Delgado, M., & Santos, y. M. (2013). El proceso de la Investigación cualitativa. Manual de procedimiento: ejemplificación con una tesis doctoral., Valladolid: Edintras.

The Informal Intercultural Mediator Nurse in Obstetrics Care

Emília Coutinho, Vitória Parreira, Brígida Martins, Cláudia Chaves
and Paula Nelas

Abstract Cultural practices linked to maternity require profound knowledge and respect for each woman and for what they consider as making sense in their experience of becoming a mother. However, the same senses are not always shared by the participants, justifying the need for intercultural mediation. The aim of this study was to understand the intercultural mediation in nursing care in obstetrics. It is a qualitative study, using the semi-structured interview to 15 obstetric nurses and analysis of the content supported by NVivo 10. Emerging categories were: meaning attributed to the intercultural mediation, principles of intercultural mediation; functions of informal intercultural mediator nurses, and reasons for intercultural mediation in obstetrics. We conclude that although in their clinical practice in obstetrics, nurses sometimes exercise informal intercultural mediator functions; they need training in intercultural mediation, which is evident from the opinions of the participants.

Keywords Intercultural mediation · To care · Obstetrics · Maternity

E. Coutinho (✉) · C. Chaves · P. Nelas
Escola Superior de Saúde de Viseu do Instituto Politécnico de Viseu, Viseu, Portugal
e-mail: ecoutinhoessv@gmail.com

C. Chaves
e-mail: claudiachaves21@gmail.com

P. Nelas
e-mail: pnelas@gmail.com

B. Martins
Enfermeira no Algarve, Algarve, Portugal
e-mail: brixida.martins@gmail.com

V. Parreira
Escola Superior de Enfermagem do Porto, Porto, Portugal
e-mail: vitvik@gmail.com

© Springer International Publishing Switzerland 2017
A.P. Costa et al. (eds.), *Computer Supported Qualitative Research*,
Studies in Systems, Decision and Control 71,
DOI 10.1007/978-3-319-43271-7_6

1   Introduction

Nurses have been undertaking the informal intercultural mediator functions in many health settings, in the relationship between the client and other professionals, and even between the client and their own imposed organizational rules. This inter-cultural mediation role has been informal; especially if we consider that the presence of formal mediators is a very recent and scarce phenomenon in Portugal.

In addition to recent (1990s decade), intercultural mediation is scarcely present in Portuguese hospitals strategy, but according to Herbert (2012) it has been an important resource for the social development of countries with a diversified cultural matrix, aimed for intercultural contact through communication. However, intercultural mediation makes sense not only when it comes to cultural diversity but also within the same family, in which each member can assign different meanings and adopt different behaviors from those expressed by the cultural norms of the group he belongs to (Campinha-Bacote 2007, 2011; Leininger 1981). Thus, even identifying cultural patterns (Leininger 1985), each individual should be cared for as a unique being, which, given their nature and life contexts, has a particular way to feel, see, think and act (Leininger 1981).

However, there is still a great lack of knowledge about the roles of nurses and their relationship with the medical practice. Nightingale's vision of the beginning of the 18th century, in which the nurse assisted the doctor in his profession, was closely linked to the biomedical model which marked the history of nurses and of the nursing profession. Dissatisfaction with the dualistic approach to the human being did raise new currents of thought. It redirected the focus of nursing practice and nursing as a human science of care has evolved and developed a humanistic perspective that is focused not in the medical practice but in human responses to health conditions, disease and life transitions, such as set in 2007 by the Order of Nurses (Ordem Enfermeiros 2007). On many different conceptual models, philosophies and theoretical models, care emerges as the essence of the discipline with the user as the center of their attention. Leininger (1970, 1981, 1985), a nurse and anthropologist, permanently influenced nursing globally and the way nurses guided their practice, making cultural care an imperative. This view holds a holistic concern that the care is significant and consistent in response to health needs and the individual nursing care, and that professionals respect the lifestyles, cultural values and family of the costumer (Leininger 1970, 1981, 1985).

In this sense, Campinha-Bacote (2011) considers it necessary and important to adopt a customer-centric approach and culturally competent in that there is a "winning and losing" by the customer and a "winning and losing" by nurses. According to the author, in addressing the cultural conflicts, or prevention, it is advisable that nurses develop the ability to distance themselves, their own cultural values and focus on the client and may, to this end, adopt different cultural skills as proposed in the LEARN mnemonic where L (listen) refers to listen to the customer, their perspective, and E (explain) refers to explain to the client what one perceived from what he said, A (acknowledge) refers to recognizing similarities and differences between the two perspectives, R (recommend) refers to recommend considering the customer's perspective, and N (negotiate) refers to negotiate with the client a treatment plan, a client-centric plan.

The maternity context is, by nature, wrapped in rituals and particular cultural practices taking on different meanings depending on the participants. In the health organizations, in the course of the interactions inherent to the care process and to the system of values, positions, interests, needs and expectations of the parties involved, which may have different views of the world, there may be constraints, which call into question the quality of care. Therefore, considering the transdisciplinary training and the fact that they continuously monitor those who need it, on a daily basis, the nurse sometimes assumes the role of informal intercultural mediator.

The care involves those who care and who are taken care of; it only makes sense when it develops beyond the technique and is guided by respect and solidarity (Polak 1996; Silveira and Fernandes 2007) by the competence with sensitivity (Herbert 2012; Polak 1996), but also by the availability (Silveira and Fernandes 2007) "attention, responsibility" (p. 79).

The essence of care in obstetrics is the act of being with the other, respecting them, demonstrating authentic presence in true interaction (Silveira and Fernandes 2007). This care allows a broader view of the environment where the woman is, (Silveira and Fernandes 2007) "her anxiety and fear she feels during the internment, the rapprochement between the nurse and the woman, non-verbal language, willingness to help, the service provided and the response that is given to every situation" (p. 79). When this environment of trust cannot be achieved, establishing a therapeutic relationship, an openness towards the other, misunderstandings can happen, as stated by Polak (1996).

Intercultural mediation is the act of a third person between two parties when, in some contexts, they cannot agree, on an issue that, due to fears, suspicions and differences, does not allow the communication between them, and who can provide the environment of trust for the parties to understand each other in order to reach a solution ACCEM (2009).

Intercultural mediation is, according to Giménez-Romero (2010), a type of intervention of a third element, oriented to the recognition of the other and to bring the parties together, "communication and mutual understanding, learning and development of coexistence, regulation of conflict and institutional adaptation, between culturally different social and institutional actors" (p. 23). The author understands that the mediator must hold some key principles required in intercultural mediation as "respect, trust, communication, language skills and intercultural competence" (p. 12). Considering the principle of impartiality the mediator cannot side with any party to advise or give instructions on what to do; he/she should only show mutual understanding and above all should not judge (Herbert 2012).

In this context, Giménez-Romero (2010) emphasizes the importance of socializing, of being with the other, being at the same time and same place as the other, with whom one interacts actively with whom one shares common features and among whom there is an understanding, empathy. For Matos (2011) it is essential to get the parties, when experiencing a mediation process, to acquire a new understanding of themselves and each other, so "they can interact more efficiently in a situation of conflict, in order to strengthen social relationships and promote the community's quality of life" (p. 23). He also understands that intercultural mediation is a voluntary process that originates from the common knowledge that objectively establishes rules of mediation, that are confidentiality, mutual respect, the recognition that "the mediator cannot suggest, decide or advise, and that the mediator should be someone who demonstrates willingness and ability for intercultural dialogue with people from different backgrounds" (p. 24).

Therefore, the aim of this study was to understand the intercultural mediation in nursing care in obstetrics.

2 Methodology

Qualitative research study, drawing on semi-structured interviews to collect data. The sample consists of 15 nurses working in obstetrics service, belonging to two districts of Portugal, one in the central region and another in the Southern Region of Portugal. The interviews took place in the period between August and October 2015, using the recording and subsequent transcription of verbatim. The guide questions were: What does informal intercultural mediation mean for you? What is your experience in informal intercultural mediation? In what context did you feel the need to resort to informal intercultural mediation? In content analysis of verbatim interviews, we used the Qualitative Analyses Software Certified Partner Program (NVivo version 10). We abided by the ethical principles with the study participants, the institutions involved and the National Data Protection Commission (Case 8981/2015, No. 9430/2015). It should be noted that 87 % of nurses do not have training in intercultural mediation.

3 Results

From the analysis of interviews to the nurses, different categories of analysis emerged: meaning of intercultural mediation, principles of intercultural mediation, functions of the informal intercultural mediator nurse, and reasons for intercultural mediation. Each table presents the category, the corresponding subcategories, the number of nurses who manifested in each subcategory (text units: TU) A number of text units greater that n means that one or more nurses mentioned more than once the same category.

For the nurses, cross-cultural mediation has different meanings, as can be seen by Table 1; finding strategies is the most referenced, for about half of the participants, followed the meaning of being an approximation. Some also mention that intercultural mediation is to find a balance and some say it is managing conflicts.

Table 1 Meaning of intercultural mediation

Category	Subcategory	n	Text units
Intercultural mediation meaning	To find strategies	7	11
	Approximation	5	7
	To find balance	1	2
	Manage conflict	1	1

The principles of intercultural mediation emerge as a category of cultural significance. Respect is appointed by two-thirds of the nurses, and the principles the most noted in this category, followed by communication, referred to by one third of the participants. Other principles are highlighted by the nurses, in particular sensitivity, impartiality, language skills, trust, the primacy of the relationship, availability, active listening, legitimation, validation and not judging the costumer, cf. Table 2.

The preventive approach to the conflict assumes greater importance in the meaning of intercultural mediation in subcategories: to inform; to understand cultural differences; and to meet individual needs, cf. Table 3. It should be noted that

Table 2 Principles of Intercultural Mediation

Category	Subcategory	n	Text units
Principles of intercultural mediation	Respect	10	17
	Communication	5	6
	Sensitivity	3	7
	Impartiality	2	4
	Language skills	2	2
	Trust	1	3
	To give primacy the relationship	1	3
	Availability	1	2
	Active listening	1	2
	Legitimation	1	1
	Validation	1	1
	Not judging the costumer	1	1

Table 3 Functions of informal intercultural mediator nurses

Category	Subcategory	n	Text units
Functions of informal intercultural mediator nurses	To inform	8	13
	To understand cultural differences	6	7
	To moderate	5	6
	To meet individual needs	3	4

Table 4 Reasons justifying the intercultural mediation in obstetrics

Category	Subcategory	n	Text units
Reasons justifying the intercultural mediation in obstetrics	Nurses understand that some women use practices less beneficial to the child's health	2	2
	The nurse is unable to satisfy the needs of the woman	2	2
	Devaluation of work in obstetrics	1	4
	Difficulty and complexity of care in obstetrics	1	3
	Nurses being transferred of service as punishment	1	3
	Nurses have an obligation to self-control emotionally	1	2
	To understand that women have to make sacrifices	1	1
	To understand that setbacks are natural	1	1
	To understand that people do not want to be counteracted	1	1
	Women hide their precarious situation for fear of removal of their child	1	1
	Women hide their financial situation due to shame	1	1
	The nurse understands that Gypsies are always transgressing institutional rules	1	1
	Imposition of professional culture	1	1

more than half of the nurses interviewed mentioned to inform as an important aspect of the functions of an informal intercultural mediator nurse and nearly half mentioned the understanding cultural differences. Direct action on the conflict is referred to by one third of the participants in the subcategory to moderate.

In clinical practice, there are sometimes situations of conflict that influence negatively the care and the relationship between nurses and clients. Despite being the category presenting lower frequencies, from the speech of some nurses, grounds for intercultural mediation emerge, cf Table 4, such as: women adopt practices that in the opinion of the nurse are not beneficial to the child's health, nurses do not adapt to the needs of women, the difficulty and complexity of care in obstetrics, service transfer of nurses as punishment, the devaluation of work in obstetrics, nurses have an obligation to self-control emotionally, to understand that women have to make sacrifices, to understand that setbacks are natural, to understand that people do not want to be counteracted, the women hide the precarious situation for fear of removal of their child, women hide their financial situation due to shame, the nurse understand that Gypsies are always transgressing institutional rules and the imposition of professional culture.

4 Discussion

Four main categories emerged from the analysis: Meaning of intercultural mediation; Principles of Intercultural Mediation; Nurse Functions while an informal mediator; and Reasons justifying the Intercultural Mediation in obstetrics.

Meaning of intercultural mediation: Just under half of participants considers intercultural mediation as finding strategies "Usually I think that any nurse arranges strategies to facilitate communication and care, basically everything that has to be solved, namely the difficulty felt" (EN), five nurses consider it as an approach "Is trying to get the culture of the women we are faced with" (ED) one defines intercultural mediation as a balance "Is trying to find a compromise between the way we think and act and the way of thinking and acting of users of other cultures" (EL), and another views intercultural mediation as managing a conflict "Often trying to mediate what we intend, or what we have to do, with what the other also intends to or not" (EJ). These findings are corroborated by Matos (2011), who defines mediation as a strategy based on communication that allows individuals to be actors and builders of consensual solutions to their conflicts, and where the conflict arises with a new connotation, as something positive, as an opportunity for change. Giménez-Romero (2010) also defines intercultural mediation as a method, with the use of a third party, geared to achieving the recognition of the other and to bring the parties together.

Principles of Intercultural Mediation: Some principles of intercultural mediation emerge from the data analysis. Respect appears as the principle of intercultural mediation referred to by most of the nurses with 17 registration units. The speeches of two nurses are highlighted, "The mediation is related to it, it has to do with us respecting the other" (EJ); "In a conflict situation... it has to do mainly with respect. If we know how the respect... we are half way through to mediation. To solve any problems that may arise resulting from these cultural differences, it goes mainly through respect" (EO). Giménez-Romero (2010) highlights respect as an important principle in intercultural mediation.

Five of the nurses interviewed, stressed the importance of communication as a principle of intercultural mediation "Anyone who in any way that facilitates communication" (EN). Also Giménez-Romero (2010) states that the communication is defined as an important principle of intercultural mediation.

Three of the nurses said it is important for nurses to have sensitivity to detect situations that need to be worked "Sometimes we have situations where the mother is very plaintive and sometimes we did not even realize that there is something else that exacerbated that complaint. And if we can get to the bottom, we can resolve the situation differently" (EL). In this context, Herbert (2012) states that in the process of mediation, it is necessary that the mediator develops a "third sense" in order to capture and decode encrypted messages, a hidden message that transmits an instruction or information on relationships.

Impartiality comes through the report of two nurses "No discrimination of users. All users have the right to be cared for with courtesy, respect, dedication and impartiality." (EN), "try to find a compromise between our way of thinking and way of doing things in the way of thinking and acting of the users of other cultures" (EL). Both the Fundación Secretariado (FSG 2007) and Herbert (2012) describe impartiality as an important feature of intercultural mediation, that the nurse should ensure it, considering it as an ally in the establishment of true mediation.

Active listening was reported by a nurse as an important principle in mediation "They have to know that there is someone who hears them, which gives them strength" (EN) "Most of the time… I try to listen to exhaustion, let the person speak all they feel" (MS). According to Morin (2002) it is important to improve the understanding we have of each other and realize that understanding is the middle and end of human communication. Domingos and Freire (2009) stress this active listening as an asset to the outcome of the conflict. Active listening is defined as an interview driving method in which the parties are on an equal footing. Through questioning and, above all, constant reformulation, "this form of listening allows the nurse to get a good grasp of the facts" (Phaneuf 2005, p. 157).

Two of the nurses consider that one of the mediating principles is to have language skills "If there is close proximity of the language, it will be easier to perceive or know something more of them to manage to provide the nursing care" (ED). One nurse said that trust is an important principle in intercultural mediation "a nurse with whom they can talk, with whom they can expose all the questions, all those more complicated situations, in the core and be able to establish a therapeutic relationship that allows this approach, in the sense that they can trust me" (EN). Trust is an emotion is the (OE 2014) "feeling of believing in goodness, strength and reliability of others" (p. 43). Herbert (2012) considers that trust is constituted as an important principle of intercultural mediation.

The therapeutic relationship is often referred to by several authors, and one nurse believes that the establishment of a therapeutic relationship is essential as a principle of intercultural mediation, with three recording units "Only if there is a therapeutic relationship they will accept what we say" (EN). Giménez-Romero (2010) also gives emphasis to the relationship, when stating that socializing is not limited to the presence of groups in a given time and place, but refers to the interaction and positive relationship between them. One of the nurses also mentioned the availability as an essential principle "Often it is not possible on the run; we have to arrange that little bit (of time) that the person needs" (EN).

Matos (2011) states that the availability must be a principle of intercultural mediation, which is understood as (OE 2014) "being prepared or available for action or progress" (p. 49). Legitimation was also identified as a principle of intercultural mediation "Not long ago I had a situation, they called me to care for a baby with trisomy 21 and the fact that I had previous experiences helped to convey a word of comfort, warmth and force. I know what it is to have in the family a child with special needs due to trisomy 21 and when I realized the situation, I managed somehow to open enough for her to accept my contact and the contact of the association trisomy 21" (EN). According to Giménez-Romero (2001), immigrants demonstrate their desire for recognition of legitimacy and competence on the part of the mediator. Validation was also reported in this category "What matters is that I realize that she understood but sometimes it is complicated because I do not speak, I speak little or no English" (EB). For Phaneuf (2005) this validation is an important mediating principle; it is a way to accept the behavior of the person and to seek it meaning (p. 543). Non judgment also arises as a principle of intercultural mediation "Above all nonjudgmental" (EH). Herbert (2012) reveals that non-judgment reflects a principle of intercultural mediation. In this sense the nurse should try not to make value judgments on the client.

Functions of informal intercultural mediator nurses: the duty to inform is the most referred function of informal intercultural mediator nurse "The key is to inform the user because even though she has a different culture, if well informed and enlightened we always end up getting the balance" (EL). Thus the nurse should be responsible not only for the care it provides, but responsible to inform the woman who is more fragile, to calm her; the nurse should be responsible, considering that woman has the right to information when it concerning her. In accordance to the OE (2014) to inform is defined as the action of "communicating something to someone" (p. 101).

Six of the interviewed nurses consider that for an informal intercultural mediator nurse it is important to understand cultural differences "Realize that not all people are like us" (EE). For Giménez-Romero (2010), it is important a mutual understanding between ethno culturally differentiated social and institutional actors. For a third of nurses, a nurse's role is to moderate "I think it's one of the nurse's role is to be the moderator of these situations" (EH), three nurses consider it to meet the individual needs "I am careful to take care of each user individually in order to meet the needs of each one, as long as I have the perception of how that user wants to be helped" (EN).

Reasons for Intercultural Mediation in obstetrics: There were several reasons, justifying the need, by the nurses, calling for the use of intercultural mediation strategies that meet the above-mentioned principles.

Women adopt practices that, in the opinion of the nurse, are not beneficial to the child's health, was one reason given by two nurses "It is still very hard to get the information through, because for a long time, it was taught to put the newborn in decubitus ventral. Afterwards they taught the lateral position on one side and the other. Due to the results of research, for many years we have been teaching to put in dorsal position, there are mothers who continue to have difficulty in accepting the supine position instead of the ventral "(EN). The nurse being unable to meet the needs of woman referred by a nurse "We sometimes do not suit to their needs and they feel needs that we cannot satisfy" (EK), and the nurses' inability to assess the satisfaction of the women's needs "we ended up not quite understanding if we managed to reach and satisfy that person the way she was expecting" (MS). The devaluation of work in obstetrics "Our peers from other areas, do not value the work of nursing in obstetrics" (EN). Bonadio et al. (2002) even refer to a sense of devaluation of midwives regarding other professionals due to non-financial recognition of their work and a certain position of submission within the team.

Situations where nurses in the past were transferred to the obstetrics service as punishment "When I started working for twenty or so years, when there was a nurse, that somehow, was in conflict with the administration, with the nursing directorate, they were placed in the puerperium" (EN). Therefore, we can understand that there was a need for mediation, for dialogue, which would lead to the resolution of these conflicts in order to avoid the punishment of nurses.

Difficulty and complexity of care in obstetrics "It is increasingly difficult. For various reasons. Because people are first time mothers increasingly later in their years" (EG), or the idea that the nurse has an obligation to self-control emotionally when he says "Professionally we ultimately strive a little more because we are working and want to present a good image to the person; in addition, we are representing a team; we cannot let down all the team" (MS).

The idea that a woman has to make sacrifices, declared by a nurse "There are circumstances that require some sacrifice, a very great gift, which sometimes causes dissatisfaction by the women because they wanted it to be all right" (EG), the setbacks are natural, referred to by the same nurse "sometimes eventually there may be some setbacks, that are natural" (EG) and yet again by the same nurse, people do not want to be counteracted, "currently they do not want to be in pain, do not want to have setbacks; They do not want to have a no, they always want yes; in obstetrics not everything is always yes" (EG).

The women hide their precarious situation for fear that their child may be removed from them. "There are situations where there is fear because they realize that their situation is very precarious, even very bad, and are afraid that somehow it means to the take the baby from them. Thus, they prefer to hide to keep their baby" (EN). Women hide their financial situation due to shame "They were unemployed, and somehow are ashamed to expose the situation, the needs they have, and they think that to resort to social worker gives a negative, and often embarrassing feeling" (EN).

To understand that gypsies are always transgressing institutional norms "There is a strong tendency to say that the gypsy culture is different, but they are keen to make a difference in their behavior, in everything; it is their culture and we have to respect it, but there are things in which not so much because they always want to be in violation of institutional rules; sometimes they come out of the visiting hours, want to stay from morning to evening" (EN). The importance of avoiding stereotypes, to non-stigmatize as regards Kleinman and Benson (2006) is emerged, justifying the need to promote dialogue and active listening in order to be able to help her overcome her fears.

The imposition of the professional culture also emerges in the speech of a participant "We also want them to respect us and it has to be us to go around them or make them understand that they also do not need to do it the way they do and maybe even they could change their habits, change something" (EE). For Rodrigues Martins and Pereira (2013) understanding and awareness of the extent of the moral and cultural aspects is required and the professional must expropriate up prejudices and stigmas that are built into the culture in order to ensure maintenance of multicultural identity and minimize the cultural imposition.

It is thus justified the need to resort to intercultural mediation and to train professionals and future health professionals for this reality. In a situation of hospitalization, of postpartum, the woman is more fragile, more sensitive and preventive intercultural mediation enables professionals to prevent conflicts, promote dialogue, demonstrating readiness and have an attitude of active listening promoting positive maternity experience.

As Kirmayer (2012) stated "there is a great need for research on the processes of implementation, clinical effectiveness, wider social impact and outcomes of culturally competent services and interventions" (p. 161).

5 Conclusion

Cultural inconsistencies arise with regard for intercultural mediation in health. On the one hand the appreciation and recognition of their nurse role with an informal mediating function, visible in categories such as significance of intercultural mediation, intercultural mediation principles, functions of the informal intercultural mediator nurse; and on the other hand the constraints experienced in some health settings that call for greater training of professionals in intercultural mediation, visible in the category reasons for intercultural mediation. We must invest more in training in intercultural mediation of nurses in order to provide opportunities for dialogue and active listening in maternity experience, which can preemptively prevent the emergence of constraints or leading to its resolution.

Acknowledgments The Portuguese Foundation for Science and Technology (FCT) through the project PEst-OE/CED/UI4016/2014, and the Centre for Studies in Education, Technologies and Health (CI&DETS); Higher Education Network for Intercultural Mediation (RESMI)

References

ACCEM. (2009). *Guía de Mediación Intercultural* Accem y Dirección General de Integración de los Inmigrantes (Ed.) (pp. 85). Retrieved from http://www.accem.es/es/guia-de-mediacion-intercultural-a725, http://www.accem.es/ficheros/documentos/pdf_publicaciones/guia_mediacion.pdf

Bonadio, I. C., Koiffman, M. D., Minakawa, M. M., & Oliveira, M. A. F. (2002). Da relação conflituosa ao respeito mútuo: a consolidação do papel da enfermeira obstétrica na assistência ao nascimento e parto. *Scielo Proceedings online.* http://www.proceedings.scielo.br/scielo.php?script=sci_arttext&pid=MSC0000000052002000100039&%20lng=en&nrm=van

Campinha-Bacote, J. (2007). *The process of cultural competence in the delivery of healthcare services: The journey continues* (5th ed.). Cincinnati, OH: Transcultural C.A.R.E. Associates.

Campinha-Bacote, J. (2011). Delivering patient-centered care in the midst of a cultural conflict: The role of cultural competence. *OJIN: The Online Journal of Issues in Nursing 16*(2), Manuscript 5.

Domingos, G., & Freire, I. (2009). Gestão de conflitos e competências da mediação informal. *Revista Galego- Portuguesa de Psicoloxía Educación, 17*(1, 2-ano 13), 85–97.

FSG. (2007). La mediación intercultural In F. S. Gitano (Ed.), Retos en los contextos multiculturales: Competencias interculturales y resolución de conflictos (Vol. 32). Madrid:

Fundación   Secretariado   Gitano.   Retrieved   from   https://www.gitanos.org/centro_ documentacion/publicaciones/fichas/23731.html.es

Giménez-Romero, C. (2001). Modelos de Mediación y su Aplicación en Mediación Intercultural. *Migraciones*(10), 59–110. doi:https://dialnet.unirioja.es/servlet/articulo?codigo=195583

Giménez-Romero, C. (2010). *Mediação Intercultural* (Alto Comissariado para a Imigração e Diálogo Intercultural Ed. Vol. 04 Cadernos de Apoio à Formação,). Lisboa: ACIDI.

Herbert, S. C. T. (2012). *Imigração, rituais e identidade: Estudo exploratório com descendentes de imigrantes cabo-verdianos* (Dissertação de mestrado), Universidade Católica Portuguesa, Lisboa. Retrieved from http://repositorio.ucp.pt/handle/10400.14/10559

Kirmayer, L. J. (2012). Rethinking cultural competence. *Transcultural Psychiatry, 49*(2), 149–164.

Kleinman, A., & Benson, P. (2006). Anthropology in the clinic: The problem of cultural competency and how to fix it. *PLOS Medicine, 3*(10).

Leininger, M. (1970). *Nursing and anthropology: Two words to blend*. New York: Wiley.

Leininger, M. (1981). Transcultural nursing: An overview. *Nursing Outlook, 32*(2), 72–73.

Leininger, M. (1985). Ethnography and ethnonursing: Models and modes of qualitative data analysis. In M. Leininger (Ed.), *Qualitative research methods in nursing*. London: Grune e Stratton.

Matos, F. A. B. (2011). *Dinamizadores Comunitários e a sua Dimensão Intercultural.* (Master), Universidade de Lisboa, Lisboa.

Morin, E. (2002). *Os sete saberes necessários à educação do futuro* (6ª ed.). São Paulo: Cortez-UNESCO.

OE. (2014). *CIPE Versão 2011 - Classificação Internacional para a Prática de Enfermagem* (O. Enfermeiros Ed.). Lisboa: Ordem Enferemiros.

Ordem Enfermeiros. (2007). Resumo Minimo de Dados e Core de Indicadores de Enfermagem para o repositório Central de dados da saúde. Ordem do Enfermeiros, 1–16.

Phaneuf, M. (2005). *Comunicação, entrevista, relação de ajuda e validação*. Loures: Lusociência.

Polak, Y. N. D. S. (1996). *A corporeidade como resgate do humano na enfermagem.* (Doutoramento), Universidade Federal de Santa Catarina, Florianópolis.

Rodrigues, F. R. A., Martins, J. J. P. A., & Pereira, M. L. D. P. (2013). Competência cultural: análise do conceito segundo a metodologia tradicional de Walker e Avant *E-REI: Revista de Estudos Interculturais do CEI*, 1–10. doi:http://hdl.handle.net/10400.22/1715

Silveira, I. P., & Fernandes, A. F. C. (2007). Conceitos da teoria humanística no cuidar obstétrico. *Rev RENE, 8*(1), 78–84.

Potentialities of Atlas.ti for Data Analysis in Qualitative Research in Nursing

Maria José Menezes Brito, Carolina da Silva Caram,
Lívia Cozer Montenegro, Lilian Cristina Rezende,
Heloiza Maria Siqueira Rennó and Flávia Regina Souza Ramos

Abstract ATLAS.ti is a program used to organize and manage qualitative research data. The present study aims at describing the application of ATLAS.ti in two instances, highlighting its potentialities and limitations for data analysis. This is a user experience of ATLAS.ti (Archivfuer Technik Lebensweltund Alltagssprache) for the analysis of qualitative data of a dissertation and a thesis in nursing. The software proved to be an efficient tool to organize, capture and analyse data; it provided the researchers with an overview of the findings during the analytical process and helped in the optimization of time. Its limitations relate to difficulties concerning characteristics of the data analysis method, the need for training and purchasing of the program. The authors concluded that ATLAS.it an important tool for data analysis in qualitative research.

Keywords Qualitative research · Data analysis · Software

M.J.M. Brito (✉)
School of Nursing, Federal University of Minas Gerais, Belo Horizonte, Brazil
e-mail: mj.brito@globo.com

C. da Silva Caram · L.C. Montenegro · L.C. Rezende · H.M.S. Rennó
Nursing Course, Federal University of São João Del Rey, São João Del Rey,
Minas Gerais, Brazil
e-mail: caram.carol@gmail.com

L.C. Montenegro
e-mail: liviacozermontene-gro@gmail.com

L.C. Rezende
e-mail: lilianc.enf@gmail.com

H.M.S. Rennó
e-mail: heloizarenno@gmail.com

F.R.S. Ramos
School of Nursing, Federal University of Santa Catarina, Florianópolis, Brazil
e-mail: fla-via.ramos@ufsc.br

© Springer International Publishing Switzerland 2017   75
A.P. Costa et al. (eds.), *Computer Supported Qualitative Research*,
Studies in Systems, Decision and Control 71,
DOI 10.1007/978-3-319-43271-7_7

1 Introduction

Qualitative researches deal with meanings, desires, aspirations, beliefs, values and attitudes, constituent elements of the human universe. The relevance of the subject and the perspective of the social players contribute to the growing demand for scientific production through funding agencies, events and specialized publications, especially in the health care field.

Qualitative research plays an important role in the latter's context as it provides methodologies and tools to approach and search people (Adorno and Castro 1994). A qualitative method enables the researcher to view the object of study, consider its conditions, its specificity and the relationships, in order to analyse and interpret it. Moreover, this approach considers meaning and intentionality as inherent to acts, relations and social structures in their beginning, transformation and in their significant human constructions (Minayo 2013).

The qualitative method stems from "human sciences" and relates to the study of a given phenomenon and the understanding of its significance to people's lives. Meaning is basic to research and, moreover, it is a method that seeks an in-depth understanding of the connection between elements which are invisible to ordinary observation (Turato 2005).

Data analysis in qualitative research needs to grasp the intentions of the research subject's discourse and capture its subjectivity. Several data analysis methods can be used: Content Analysis; Textual Analysis and Discourse Analysis. The process of data analysis, regardless of the method chosen, involves long, complex and intense handmade techniques for the construction and deconstruction of texts in order to capture the essence of the subjects' statements. This process requires the researchers' time and depends on their organization skills so that the material is fully used and data loss is avoided.

In recent years, efforts have been made to develop data analysis programs that support and accelerate analytical processing, as well as enable new ways of doing things. Given the variety of data that can be used to the interpretation of social phenomena, researchers seek tools that can support data organization and management.

ATLAS.ti has been used as an operational tool during the analytical stage of nursing research on the interpretation of social and human phenomena. Such studies seek to understand the meanings attributed by people and to examine the subjective understanding of phenomena (Trindade et al. 2013; Silva et al. 2010; Amaral et al. 2012; Ramos et al. 2013). The present paper aims at examining the ATLAS.ti software considering the results of qualitative researches and examining its potential use as a data analysis tool.

ATLAS.ti was developed at the University of Berlin in 1994 and it has been applied to the analysis of qualitative data in various areas (Walter and Bach 2009). The acronym "ATLAS" stands for Archivfuer Technik Lebensweltund Alltagssprache in German, or "Archive for Technology Lifeworld, and Everyday Language"; "ti" means text interpretation (Bandeira-de-mello and Cunha 2003). It is known as well as CAQDAS (Computer Assisted Qualitative Data Analysis Software) which refers to that program use as qualitative data analysis software.

The software has four guiding principles of analysis (Bandeira-de-mello and Cunha 2003): visualization, which allows the researcher to manage the complexities of the review process, organizing and maintaining continuous contact with data; integration of database and elements, ensuring the contact of the researcher with organized data; randomness, which promotes the discovery and insights casually, that is, without a deliberate search; and the manipulation of the material, i.e. the interaction between the various constituent elements of the program, which promote new discoveries and insights.

It is important to point out that the program is a resource, a facilitator, not an end in itself (Teixeira and Becker 2001): a computer program will never replace the researcher's creativity, common sense and sociological perspective. Therefore, the choice of the software should be associated with the object of study and research approach, so that its use is consistent and subordinated to the researcher's perspective (Klüber 2014).

Considering the above, the objective of this study was to describe experiences on the application of the ATLAS.ti, highlighting its potentialities and limitations for data analysis.

2 Methodology

This is a user experience study on the applications of ATLAS.ti for the analysis of two qualitative research data: a dissertation and a thesis in nursing completed in December 2015.

The dissertation aimed at the development of ethical and moral competencies, as well experiences of moral distress amongst nursing students (Rennó 2015). The study participants were 58 undergraduate nursing students at two Higher Education Federal Institutions, in their last year course which is the internship. Data was collected during eight focus groups comprising six to eleven students per group; a secondary data collection was carried out through the analysis of the Political and Pedagogical Project (PPP) of the courses. Data were analysed using Thematic Content Analysis through ATLAS.ti 7.0 (Bardin 2011).

The thesis studied comprehensiveness in the practice of professionals of Family Health Strategy (ESF) teams in the context of Primary Health Care (APS). The study participants were five ESF professionals and 14 registered users and chosen by data saturation (Turato 2005). Data was collected through observation, semi-structured interviews and techniques using comic books (Rezende 2015). The statements of interviewees and participants in comic books technique were recorded and transcribed in full in order to preserve its reliability and integrity. Data was analysed via Thematic Content Analysis (Bardin 2011) using ATLAS.ti 7.0 resources.

According to Bardin (2011), thematic content analysis consists of a set of analysis of interactions between researcher and research participants, aiming at identifying the essence of the latter discourses in order to describe the content of their messages. It is a method that aims at overcoming the common sense and the subjectivism of interpretation and performing a critical analysis of texts, biographies, observations or interviews.

In general terms, a thematic content analysis relates semantic structures (significant) and sociological structures (meanings) of statements, articulates the descriptions and analysis of texts with the factors that determine their characteristics, which are: psychosocial variables, cultural context and message production process (Bardin 2011). In this sense, by having as object the individual speech and its language, the method aims at understanding subjects in their own context at any given time, seeking to grasp their reality of life through the words of their discourse (Bardin 2011).

The researchers carried out thematic content analysis following its three stages: (1) pre-analysis; (2) exploration of the material and; (3) inference and interpretation (Bardin 2011). Those stages will be developed along this article, since they are essential to the understanding of ATLAS.ti application and the highlighting of its potentialities and limitations in data analysis.

3   Development: Application of ATLAS.ti in Thematic Content Analysis

The authors will identify content analysis stages and describe the use of the program in each of them. Figure 1 below shows the layout of the software and the main tools used in the analysis. Those are presented in two columns: left column shows the material to be analysed; and right column displays the codes derived from quotations.

3.1   1st Stage: Pre-analysis

In this stage, the material in text format was saved on the computer and inserted into the ATLAS.ti template, composing for each search a hermeneutic unit. In the case of the thesis the texts inserted into the program were: political and pedagogical projects of the courses; transcribed interviews carried out in the focus groups. In the case of the dissertation, those were: field diaries observations; transcribed individual interviews.

After insertion of the documents, the researcher performed an exhaustive reading (first reading of the document that allows the researcher to be invaded by impressions and guidelines) of interviews and documents seeking to identify,

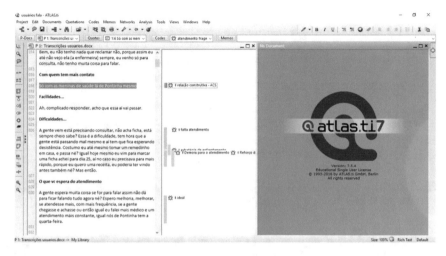

Fig. 1 ATLAS.ti presentation page

analyse and define the corpus. The corpus was formed through the break-up of texts in units and their grouping by similarity—delimitation of material (Bardin 2011).

The inclusion of the material in the program and the creation of hermeneutic units enabled data visualization and organization. The software allows the grouping of materials in different formats such as Word, PDF and JPEG. Furthermore, by allowing the inclusion of multiple documents, data from different surveys can be grouped to compose a corpus.

3.2 2nd Stage: Data Exploration

It consisted of breaking the text in codes (coding) and categorizing them.

During coding, a correspondence between the content of the documents (corpus) and the identified units (recording unit) was performed. Subsequently, the researchers identified register units by a segment of the message that gave it meaning, i.e., composing the context units, reaching the representation of the content and expression. In ATLAS.ti, register units are called quotations and context units are codes.

Codes were created and re-elaborated countless times in order to achieve the research objectives. At this point, the researchers realized the agility and convenience of the software for the organization and management of the material without the use of pen and highlighters: the researchers coded the corpus and selected sentences and/or words, given that the program allows the researcher to retrace the path of analysis whenever deemed necessary. The program allows as well the creation of research notes and comments in the codes.

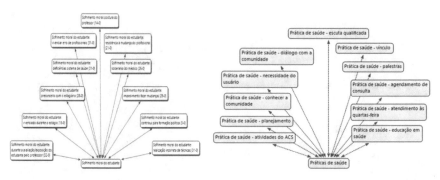

Fig. 2 Code network related to family: moral distress of students (fig *left*—doctorate) and health practices (fig *Right*—master's) generated by ATLAS.ti

The categorization stage defines results; it is based on two main criteria—repetition and relevance—and intends to create a representation of the data which will then be interpreted (Turato 2005). In ATLAS.ti, categorization was performed through organization of the codes according to their common characteristics or relevance, generating what is called Family.

A Fig. 2 shows a graph generated by the program of a Family with their respective codes to each search. The Family can be grouped and regrouped, creating thematic categories. The doctoral dissertation generated 392 codes and 47 Family. The Master's thesis generated 93 codes and 08 Family.

The categories generated in the doctoral dissertation were "moral suffering and facing moral problems during Nursing education"; and "Different perspectives on ethical and moral skills in nursing education." In the master's thesis, those were: "The daily life in the Quilombo Community of Pontinha"; "Characterization of Daily Living Practices"; "Care delivery in the perspective of ESF professionals"; and "Perception of Users on the ESF Daily Practice: between healthcare needs and access to health care."

One of the limitations of the program is the automatic junction of Family, since grouping is performed by the researcher, off-program. The researcher selects the Family individually for the formation of categories. Experience suggests that manipulation of the Family should be at a later stage of the process, especially considering the content analysis that requires such organization.

Another limitation relates to thematic content analysis, refers to exclusivity rule. The exclusivity rule proposed by Bardin (2011) means that a same quotation cannot compose different Family. In the operationalization of Family, the codes representing the quotations are available for use, even after being part of a Family. Such situation confuses the researcher, since it is not clear which quotation was used in each family. The issue is only recognised when the Family is already made. So it is up to the researcher to be aware of the grouping of codes.

3.3 3rd Stage: Inference and Interpretation

In this stage categorized data will be processed in order to have meaning, to deepen the analysis and give more density to the research (Bardin 2011). The results therefore established a dialogue with the literature.

At this stage, data has been properly managed in ATLAS.ti. Its application, at this point, enables the viewing of codes present in each Family, in which reference quotations can be found. This contributes to interpretation and discussion. Another facilitator is the export of data created in the program in different formats (DOC, XML, HTML, SPSS). The software enables the researcher to organize data. Its effectiveness depends on the researcher's ability to manipulate it in search of its object of study and to relate it to the theoretical framework.

In the thesis, the aspects that gave support to the analysis of the focus groups addressed issues related to the moral distress of nursing students, emphasising internship and coping strategies. Its analysis included aspects of Political and Pedagogical Projects that could contribute to the development of the student's ethical and moral skills, like school records, legal bases, general guidelines of the course, curriculum aspects, especially internship and general competencies.

In the dissertation, aspects that supported the analysis addressed health practices developed by the professionals, the user's experience in the APS and their health needs, focusing on the comprehensive practices recommended by the Unified Health System (SUS). Observations made by the researcher constituted the corpus.

The software made it possible to work with non-numeric and unstructured data, favouring their indexing, searching and theorizing. Thus, the results were treated as to become significant.

4 Impressions on ATLAS.ti

ATLAS.ti software allows data organization, management and visualization in qualitative research. It deals with large amounts of text, in various formats (word, the richtext and PDF) and types, such as text (answers to non-structured questionnaires, transcribing interviews, observation reports, documents, letters, journalistic or literary text and others); audios (interviews, meetings, music and others); images (photos, drawings, paintings and others) and videos (of semi-experimental research, films, television reports and others) (Walter and Bach 2009). Therefore, it is a flexible, applicable and useful tool that works in different formats and adjusts to data, objectives and research strategy.

It presents advantages in data structuring as it allows the researcher's contact with and focus on the original data, revealing hidden patterns and powerful insights through the construction of semantic networks, sophisticated searches, charts and graphs. The program can store the records of all the stages of analysis, as well as it allows the researcher to retrace the organization of the material when deemed necessary (Walter and Bach 2009).

ATLAS.ti contributes to the automation of the analysis process, since it computerizes the research files. It facilitates the operationalization of the research and assists in deep data analysis, enabling the location of multi-data simultaneously (Walter and Bach 2009). ATLAS.ti saves time, because data analysis is faster; it generates rich, in-depth analysis, without data loss and with crossing of reliable data (Teixeira and Becker 2001). Researchers should, however, remain constantly observant, their focus should be on the object of study and they should always be creativity (Teixeira and Becker 2001).

The program is a resource, a facilitator not an end in itself; its application cannot replace creativity, common sense and the researcher's sociological perspective (Klüber 2014). It is worth emphasizing that a theoretical framework for data analysis should be used in line with the study objectives and the approach proposed by the instrument to be used.

One of the software limitations is the need of training to handle it. Although it has a user friendly template, it is little self-explanatory and not available in Portuguese. In addition, it requires purchasing of a license, which limits the number of users.

The studies used in this research identified ATLAS.ti as an appropriate tool for data organization in thematic content analysis.

5 Conclusions

This study aimed at describing the strengths and limitations of ATLAS.ti 7.0, defining its use in a doctoral research and master's degree in nursing.

Data analysis used content analysis and was organized by ATLAS.ti. The software was used as an operational tool to optimize the management of information by organizing collected data, favouring indexing, searching, theorizing and qualification of the findings. The software is an excellent analytical tool but, to be used successfully, it requires all stages of the research to be performed with excellence using an appropriate theoretical framework and a clear delineation of the research problem and goals.

ATLAS.ti creates Family and Codes groups in a process of construction and reconstruction, until the proposed objectives are achieved. In this sense, one of the potential uses of ATLAS.ti software was to support the researcher's process of making and remaking the data coding path.

The program contributed, as well, to data organization regardless of their volume. Furthermore, it allowed the concomitant development of other researches, expanding the use of software as essential tools that represent real time and quality gains, generating rich and in-depth analysis through data crossing.

The research identified as a limitation the need for training, since the program is not a self-explanatory instrument. Therefore, it is important that the researchers involved in the analytical process are trained to use the software, so that the program's advantages are fully used. Accessibility is hampered by the need to purchase a license and its unavailability in the Portuguese language. The present research demonstrates that further studies involving the use of ATLAS.ti should be carried out so that its limitations and potentialities are revealed.

Acknowledgments The researchers would like to thank the Administration for Nursing Research Centre (NUPAE-UFMG); the Study Group on Work, Citizenship, Health and Nursing (PRAXIS-UFSC); the CNPq and CAPES for financial support and scholarship.

References

Adorno, R. C. F., & Castro, A. L. (1994). The exercise of sensitivity: qualitative research and health as quality. O exercício da sensibilidade: pesquisa qualitativa e a saúde como qualidade. *Saúde Soci-edade, 3*(2), 172–185.

Amaral, R. F. C., Souza, T., Melo, T. A. P., & Ramos, F. R. S. (2012). Therapeutic itinerary in the mother-child care: interfaces between culture and biomedicine. Itinerário terapêutico no cuidado mãe-filho: interfaces entre a cultura e biomedicina. *Revista Rene, 1*(13), 85–93.

Bandeira-de-mello, R., Cunha, J. C. A. (2003). Operationalizing the method of grounded theory in re-search strategy: technique sandanalysis procedures with the support of Atlas/ti software. Opera-cionalizando o método da grounded theory nas pesquisas em estratégia: técnicas e procedimentos de análise com apoio do software Atlas/ti. In Encontro de Estudos em Estratégia, Rio de Janeiro: ANPAD.

Bardin, L. (2011). Análise de conteúdo. *Edições, 70*, Lisboa.

Brito, M. J., Ramos, F. R. S., Caram, C. S., Caçador, B. S. (2014). Ensino de Administração em Enferma-gem: o olhar dos protagonistas que vivenciam o processo de aprendizagem. In M. J. Brito, F. R. S. Ra-mos, C. S. Caram & B. S. Caçador (Eds.), Administração em Enfermagem-estratégias de ensino. Coopmed (pp. 163–174).

Justicia, J. M. (2003). *Análisis cualitativos de datos textuales con Atlas TI*. Espanha: Universidade Autônoma de Barcelona, Versión.

Klüber, T. E. (2014). ATLAS.ti as an analytical tool in qualitative research of phenomenological ap-proach.Atlas.TI como instrumento de análise em pesquisa qualitativa de abordagem fenomenoló-gica. *Educação tematática Digital, 16*(1), 5–23.

Minayo, M. C. S. (2013). The challenge of knowledge: qualitative health research. O desafio do co-nhecimento: pesquisa qualitativa em saúde (8th ed.). São Paulo: Editora Hucitec.

Ramos, F. R. S., Brehmer, L. C. F., Vargas, M. A. O., Schneider, D. G., & Drago, L. C. (2013). The ethics that is built on nursing education process: concepts, spaces and strategies. A ética que se constrói no processo de formação de enfermeiros: concepções, espaços e estratégias. *Revista Latino-Americana de Enfermagem, 21*, 113–121.

Rennó, H. M. S. (2015). *Desenvolvimento de competências ético-morais e o sofrimento moral na for-mação em enfermagem*. 180 p. Tese (Doutorado em Enfermagem). Programa de Pós-Gradua-ção em Enfermagem, Universidade Federal de Minas Gerais.

Rezende, L. C. (2015). *O cotidiano de uma comunidade quilombola: a (des)construção da integrali-dade na visão de moradores e equipe de saúde*. 109 f. (dissertação de mestrado). Programa de Pós- Graduação em Enfermagem, Universidade Federal de Minas Gerais.

Silva, D. G. V., Souza, S. S., Trentini, M., Bonetti, A., & Mattosinho, M. M. S. (2010). The challenges fa-ced by beginners in nursing practice. Os desafios enfrentados pelos iniciantes na prática de en-fermagem. *Revista da Escola de Enfermagem USP, 2*(44), 511–516.

Teixeira, A. N., & Becker, F. (2001). New possibilities of qualitative research via CAQDAS systems. Novas possibilidades da pesquisa qualitativa via sistemas CAQDAS. *Sociologias*, Available from http://dx.doi.org/10.1590/S1517-45222001000100006

Trindade, L. L., Lima, L., & Pires, D. E. P. (2013). Implications of care models of primary care in the workloads of health professionals. Implicações dos modelos assistenciais da atenção básica nas cargas de trabalho dos profissionais de saúde. *Texto & Contexto Enfermagem, 1*(22), 36–42.

Turato, E. R. (2005). Qualitative and quantitative methods in health: definitions, differences and re-search objectives. Métodos qualitativos e quantitativos na área da saúde: definições, diferenças e seus objetivos de pesquisa. *Revista de Saúde Pública, 39*(3), 507–514.

Walter, S. A., & Bach, T. M. (2009). Goodbyepaper, highlighters, scissorsandglue: innovatingtheproces-sof contente analysis through Atlas.ti. Adeus papel, marca-textos, tesoura e cola: inovando o pro-cesso de análise de conteúdo por meio do Atlas.ti. In Seminários de Empreendedorismo e Edu-cação, São Paulo: Universidade de São Paulo (USP).

Computer Assisted Qualitative Data Analysis Software. Using the NVivo and Atlas.ti in the Research Projects Based on the Methodology of Grounded Theory

Jakub Niedbalski and Izabela Ślęzak

Abstract Our presentation raises the matter of the application of specialist software which assists qualitative data analysis in research based on the procedures of grounded theory (GT) methodology. The aim is to present the relationships between the procedures of GT methodology and the Atlas.ti and NVivo software. During our presentation, we would like to demonstrate the manner in which the functions available in the Atlas.ti and NVivo software may be applied when carrying out analysis based on GT methodology. Hence, while we focus on technical issues (collecting, editing, segregating and ordering the data), we also look at the analytical possibilities (the process of coding, searching in respect of codes, writing memos, creating relationships between the codes, establishing a network view) of the presented software.

Keywords Grounded theory methodology · CAQDAS · Atlas.ti · Nvivo

1 Introduction

Development of modern technologies provides researchers with new manners and opportunities of realization of research projects. Quickly developing computerization and informatization bear great significance in this field. Modern technologies impact the research process increasingly more intensively, by providing innovative methodological tools, such as specialist computer programs. Within recent decades, we have been able to notice highly dynamic development of computer-assisted qualitative data analysis software, and the list of available programs is constantly becoming longer (c.f Lewins and Silver 2004). Apart from relatively simple tools with limited possibilities, we have so developed programs as Atlas.ti or NVivo,

J. Niedbalski (✉) · I. Ślęzak
Faculty of Economics and Sociology, Sociology of Organization
and Management Department, Institute of Sociology, Lodz University,
Rewolucji 1905r. Nr 41/43, 90-214 Lodz, Poland
e-mail: jakub.niedbalski@gmail.com

© Springer International Publishing Switzerland 2017   85
A.P. Costa et al. (eds.), *Computer Supported Qualitative Research*,
Studies in Systems, Decision and Control 71,
DOI 10.1007/978-3-319-43271-7_8

which provide the researcher with possibilities to establish connection between codes, to perform complex searches for data, generate hypotheses, and further also to construct theories (Fielding 2007: 463; Kelle 2005: 486). Such software helps in the creation of complex collections of data, and their comprehensive arrangement, according to the researcher's intentions. NVivo and Atlas.ti have a quite long history now, and they managed to reach leading positions among CAQDA software. These are tools that underwent numerous modifications, and which have been continuously improved for more than twenty five years, paving the way to quality researchers.

In our case, we used opportunities provided by both these programs to analyze the data on the basis of grounded theory methodology. Our comments and observations related to the process, are briefly described below.

2 Background

In our research we attempted to identify the perspective of the examined social actors, understanding the procedural dimension of the researched phenomena concerning people with disabilities who are socially excluded. Therefore, the grounded theory methodology turned out to be especially useful for us. We were aware that this choice of methodology is connected with particular ontological and epistemological assumptions, which form the framework of the research conducted. Hence, we made every effort for the research process to take place in accordance with methodologically correct procedures which, in the case of GT, comprise open, selective and theoretical coding, theoretical memos, theoretical sampling or a constant comparative method.

Taking the complexity of the GT procedures into consideration, together with a significant number of data, which we were keen to analyse, we started to look for a manner for rendering the research process more efficiently and effectively. We came to the conclusion that a good solution would be to find support in the form of computer programs (CAQDA). Therefore, after a trial-and-error period, we selected the Atlas.ti program, which we believe suits our needs best.

3 Methods

Grounded theory is treated in the literature of the subject as a research strategy, the main purpose of which is to develop a theory (c.f Creswell 1998). This strategy consists in development of a theory (of a moderate reach) based on systematically collected empirical data (Strauss and Glaser 1967; Glaser 1978). Therefore, the theory derives from analyses of empirical data, showing itself during systematic field research, arising from data, which are directly referred to the observed part of the social reality. Hypotheses, notions and their qualities are developed, modified and verified during empirical studies. Therefore, development of a theory is strictly connected with a long-standing research process. It may be stated that a researcher's purpose is gradual transfer from empirical material to higher levels of abstractive thinking, through creation of hierarchically differentiated categories and their qualities, to the construction of hypotheses and theories.

The logics of the research process is based on seeking an increasingly higher conceptual level, and as a result, dropping data and turning to theorization. A key role is played in this context by the process of coding, i.e. ascribing batches of material with particular labels that reflect their sense and meaning allocated by social actors, and reflected by the researcher. These actions are accompanied by particular procedures of methodological correctness, which in the case of GT include, among others, theoretical sampling, constant comparative method, coding, writing notes. The process of empirical data collection takes place, in the case of the grounded theory method, gradually, on various phases, but alternately, with analysis and interpretation performed in parallel.

4 Working Environment

Both presented programs are powerful tools aiding the qualitative analysis of text data, as well as of graphic materials and audio/video files. NVivo and Atlas.ti offer an array of functions supporting tasks connected with a systematic approach towards qualitative data analysis. Atlas.ti and NVivo support the process of exploring phenomena hidden in the data, and they help to deal with the complexity of analytical procedures, at the same time offering a friendly and intuitive working environment. Both programmes support researchers' concentration on material analysis, providing functions facilitating management, edition, comparing and creating hypotheses and theories out of significant loads of data, all of this in a creative, flexible and at the same time systematic way (Fielding and Lee 1991).

It is worth emphasizing that the discussed programs belong to the category of applications supporting development of theories, and this is why they correspond very well to the requirements put forward by the grounded theory methodology procedures. This in turn generates specific solutions implemented in NVivo and Atlas.ti, and which are directly reflected in the appearance of the applications themselves, their structure and available options.

The creators of both programs face various habits of analysts—as ordinary computer users—who have their customs related to the manner in which they operate the equipment and particular software. NVivo allows, among others, to show, hide and move certain elements visible on the screen. There is also a possibility to use certain functions from the drop-down menu, or icons located on the menu bar, and through certain combinations of keyboard shortcuts (Schönfelder 2011). As a result, the program offers significant freedom in creating the working space, and adjusting it to individual user's needs. The case is similar with the Atlas. ti program, where users may implement various modifications in an easy manner, according to their preferences. The HU editor serves as the main instrument, which offers access to all other tools of the program. Users may decide which parts of the window to display and adjust its appearance to their own needs.

4.1   Open Coding

A basic analytical action that we undertook was the coding process, which consisted of ascribing data fragments with labels describing the content of each particular fragment. The coding took place according to the open scheme, when numerous codes were ascribed. Afterwards, we moved to axial coding—which focused on selected key categories of a higher, theoretical meaning (Strauss and Corbin 1990). Such coding performed in Atlas.ti and NVivo took place through marking a text, fragment by fragment, with each section subsequently being ascribed certain labels. What is significant is that as our ideas evolved the generated codes could be modified. Therefore, a previously ascribed code could change its name, become merged or replaced with another code, as the theorization process progressed.

During the analysis process, we developed almost 500 codes, which were gradually combined, grouped or simply excluded, as a consequence providing a number of 40 categories. While the interviews and notes from observations were coded in two manners: "in terms of scope", regarding the time, when the story took place, and "in terms of details", where certain fragments corresponding to particular leads were marked in relation to scopes. A key of material codes was developed during the analysis. Codes grouping into a hierarchical code took place from the bottom. Each period was provided with almost a hundred codes, which were afterwards arranged into more general categories, until three main groups were achieved.

It seems that authors of NVivo and Atlas.ti assumed a possibility of coding as one of the main objectives while developing the software. From this perspective, the NVivo and Atlas.ti applications seem to meet the researchers' expectations, who use the grounded theory methodology in their research, especially that coding and categories generation, and creation of connections between them, pose a major part of work done by an analyst using GT. A result of categorization and combination of codes is the construction of hypotheses and development of theories on their basis.

Both programs, Atlas.ti and NVivo proved to have similar functionalities in the process. Both allowed to carry out the coding process, which, according to us, remained in compliance with grounded theory methodology procedures.

4.2 Relationships Between the Codes

Another step in the analysis based on GT is axial coding, when connections between categories and subcategories were determined. Analytical procedures of focused coding allow, according to Anselm Strauss, to recognize the relationships between the structure and the process. Also in this scope, the NVivo and Atlas.ti programs provide the researcher with support, because they have elaborated functions, which support development of the hierarchy of categories, but also enable the creation of connections between codes. It is significant because thanks to such tools, it becomes possible to continue the analysis and to take it to a higher conceptual level (Fielding and Lee 1998). The existence of various functions, which on one hand allow to create a structure of categories, and thus facilitate the process of arranging the coding results, therefore enabling reflection of the "superiority-inferiority" relation (including specification of categories, subcategories, and their qualities), and on the other hand allow to determine the character and relations connecting the generated categories more precisely, cause it to become real to use the coding paradigm in the coding process. In other words, the coding paradigm means a general theory of action, which may be used to develop a structure or "axis" of a developing theory.

In the research that we described, we applied the five-element of Strauss and Corbin model (1990), in order to construct an initial analytical scheme: causative conditions, intervening conditions, context, micro-actions and consequences. Using the suggestions proposed by Strauss and Corbin (1990), we constructed a structure of notions explaining the phenomenon that we studied. After completion of those actions, the purpose of studies, which was initial and general until now, has become more concrete and precise. Using the suggestions proposed by Strauss and Corbin (1990), we constructed the following structure of notions explaining the phenomenon that we studied:

- First of all, the causative conditions, which in our study meant the level of motivation to act despite the disability;
- Second of all, the phenomenon (the main category), so, according to our analyses, it was the reconstruction of the process of going out of the social space of marginalization and social exclusion;
- Third of all, the intervening actions, meaning the cognitive model or the concept of disability;
- Fourth of all, the context, which was brought down, among others, to limitation or strengthening of environmental impacts, as well as actions and behaviors of other people;
- Fifth of all, strategies of actions/interactions, represented by two extreme categories: (a) independence (inner locus of control), (b) dependence (outer locus of control).

Moreover, on the basis of the coding paradigm's elements determined as such, we managed to operationalize the so-called consequences, which meant the reconstruction of the notion of disability, and the assumption of the concept of social exclusion of this category of persons.

4.3   Searching with Regard to Codes

The logics of the research process in grounded theory methodology is based on seeking an increasingly higher conceptual level and, as a result, dropping data and turning to theorization. In this context, the constant comparative method, which consists of searching for differences and similarities between fragments of data, codes or cases is of great significance. Increasingly, more general categories revealing underlying uniformities are generated on the basis of similarities and differences' analysis. Preservation of a work style, recommended by the authors of *The discovery of grounded theory*, i.e. treating the study as a whole process that is consciously directed at generation of theories, leads very quickly to—as claimed by Strauss and Glaser (1967)—the formulation of a multiplicity of hypotheses. At the beginning not connected, in a short time they start to form a theoretical framework of the research. Therefore, from the perspective of the processual character of theories generation, it seems useful that the created inquiry and interconnected search query may be updated according to subsequent changes introduced into the project. We can restate the inquiry in regular time intervals, and hence monitor the development process of our coding, and evaluate whether the recent analyses head in a direction that is satisfactory for us.

Computer programs, which we used to analyze the data, have special functions, which enabled to "verify" the hypotheses through scanning parts of interviews and notes from observations. Therefore, computer assisted qualitative data analysis software may be useful to improve theoretical concepts, and to create and "establish" hypotheses.

In our research, we verified our intuitions related to the impact of the process of leaving social marginalization of a disabled individual, exerted by social welfare institutions. In order to research such an initial hypothesis, the option of searching data in terms of spatial presence of codes in source materials turned out to be useful. For example, we can introduce proper configuration, pointing to segments of texts coded with the first code, and segments coded with the second code, appearing at a certain distance from the first ones. Therefore, the hypothesis of relation between leaving social marginalization and impact by a social welfare institutions, may be studied through searching for all elements of the text, coded by "leaving from social marginalization", and fragments coded with "impact by a social welfare institution", located at a certain distance from the first one (expressed with a number of verses).

In practice, while using Atlas.ti or NVivo software, the procedure of comparison was realized through the application of the data search option. This process consisted of reviewing fragments of a text and other data, which were coded with a particular code. Hence, we obtained knowledge on opinions of a given topic, which we understood in an analytical category, expressed by particular speakers. Also in this scope, both programs showed similar functionalities, although in our opinion, NVivo turned out to be more intuitive. Atlas.ti required us to spend more time to get familiar with its functions of searching through data.

4.4 Memos

A crucial action within the process of the generation of theories was theoretical coding. It takes place through theoretical memos, i.e. the thoughts of a researcher about the encoded categories and hypotheses, written in a theoretical language. Notes writing has accompanied an analyst applying the grounded theory methodology since the beginning of the research process. These notes may be related to the whole project, collected data in general or any source of data individually, and also subsequent stages of analysis and particular codes (Saillard 2011). Therefore, preparation of notes is of crucial meaning on each level of the coding and data analysis process (c.f Strauss and Corbin 1990). In other words, starting with open coding, a research should, as far as possible, write down all ideas related to the process of interpreting data and drawing conclusions from the analysis. In this context, the tools, e.g. comments, notes or annotations—offered by CAQDAS packages—gain on importance for the qualitative approach, enabling realization of analyses with application of the grounded theory methodology. Therefore, Saillard suggests to call them "reflection tools" of a researcher (2011).

In both programs, *memos* are separate components of the whole project, which could be connected with other elements, such as codes or source documents. In the NVivo and Atlas.ti programs, a role of theoretical notes is played by memos, i.e. records of theoretical thoughts and concepts by a researcher. The concept of memos in the discussed programs is analogical to the procedure of generating notes in the grounded theory methodology. They are there to help the researcher move to a higher conceptual level, and they serve the generation of theories as tools of theoretical coding. Also in this case, a researcher gains a possibility to ordinate such data, and to provide them with a certain structure, e.g. by cataloging memos, and describing them with data, in order to monitor the changes arising in the process of data analysis and interpretation. Writing notes is a significant task on each stage of qualitative analysis. The ideas recognized in the notes are often certain "puzzle pieces" which can be connected afterwards and used at the stage of writing reports. Both programs enable the researcher to fluently move between open and focused coding, writing memos, modeling. Through that, they can support the process of collecting, analyzing data and theorizing (Bringer et al. 2006).

4.5   A Mind Map

Strauss and Corbin (1990) suggest that during the process of generating theories, apart from coding, segregating and arranging information, seeking for patterns between data, using a system of notes, we should also lean on visual representations of connections and interdependencies between generated analytical categories. Any visualizations in the form of charts, diagrams or networks are useful to organize relationships between categories, which emerge during selective coding. From the GT methodology perspective, the most significant are the models, which form the basis for diagrams that integrate data. Any schemes, diagrams or models are used for visual representation of connections and interdependencies, which exist between components of the developed theory. What is more, contrary to linear representations of various relationships, the network distribution of those relations is closer to a human manner of perceiving reality, and it therefore becomes one of the most important interpretation processes of an analyst.

In the Atlas.ti and NVivo software, we used the function 'network view creation' to determine and review initial concepts and ideas on questions which lie in the field of our interest. We also used it to create visual representations of relationships between elements of the project, to identify emerging patterns, theories and explanations, or document and register subsequent stages of work over the project.

Thanks to such a visualization of data, it has become much more convenient for us—researchers performing analyses that head towards generation of theories—to compare various elements of a single project. First of all, representation of subsequent stages of an analysis in the form of a mind map allowed improved observation of relationships and patterns of data. At the same time, application of the modeling function allowed to develop a project draft, and a vision of our own ideas as related to the development of the material.

At the end, it is worth adding that while using a computer program, and creating various visual representations of the data analysis that we led (including creation of an integrating diagram that somehow crowned the whole process), we can document the course of all actions undertaken by a researcher in this scope. Therefore, we—as researchers—can monitor subsequent stages of formulation of our analytical "path", and we also become more transparent for our recipients thanks to that, who obtain a possibility to look into the course of the whole research process. Such a direct and tangible expression of this action may be the presentation of a natural history of the study, which we can not only present in a descriptive manner, but also through data exported from the program, and include them into our work in the form of a report.

5 Conclusions

Atlas.ti and NVivo may be highly supportive in every scientific field, and they may have highly practical application whenever qualitative data is used. NVivo an Atlas.ti software are equipped with instruments which facilitate the process of meeting the requirements connected with generating grounded theory, providing the researcher with new tools allowing to take care of the emerging theory to be fit and modifiable (Glaser 1978). What makes NVivo and Atlas.ti interesting for analytics applying grounded theory methodology is the possibility to develop material according to the logics of abductive research conduct. It means that the researcher becomes equipped with such functions and options of software which allow to follow the path leading from detailed data to developing general conclusions that may then be verified by the researcher coming back to the source materials.

During work in the described programs, while using the implemented functions, we managed to perform data analysis (interviews, notes from observations of existing data, as well as audio and video materials), in a manner which corresponded to requirements put forward by grounded theory methodology procedures. It is also worth emphasizing that the programs turned out to be helpful as tools for collection of, and at the same time control over a significant amount of materials, which may be simply processed, modified, sorted and reorganized, as well as sought through. It enables a researcher to gain greater control over the collected data. It is also accompanied by a possibility to subordinate various elements of the projects, among others, through grouping them in accordance with the preferences of a researcher (Wiltshier 2011). The Atals.ti and NVivo programs allow comprehensive arrangement of data—both source materials and any information resulting from an analysis carried out by a researcher. It also must be kept in mind that the computer software devoted to qualitative data analysis creates a possibility of constant modification of all project elements, together with the emergence of new data (Bringer et al. 2006). A flexible manner of creating and modifying the project elements allows a researcher to follow the data, and the generated code may be quickly modified if we decide that it does not reflect the data content to a sufficient extent (Glaser 1978). At the same time, the system of analytical notes allows fluent alteration of actions related to collection and analysis of data.

From that view, Atlas.ti and NVivo programs allow to conduct particular tasks regarding the analytical process, which would be impossible in such a dimension or in such a relatively short time when applying traditional ways of research. Nevertheless, it needs to be noticed that computer aided qualitative data analysis is not equal to the best method for designing and conducting research, but it is a kind of alternative for traditional methods. The choice that will be made by the researcher should depend on his/her personal preferences, type of research and character of the explored field. It seems that the most important aspect is not about the way of carrying out the research—traditional or with CAQDA—, but about the question of choosing proper research techniques, methods and tools that will be adjusted to the actions planned by the researcher (Kelle 1995).

References

Bringer, J. D., Johnston, L. H., & Brackenridge, C. H. (2006). Using computer-assisted qualitative data analysis software to develop a grounded theory project. *Field Methods, 18*(3), 245–266.

Creswell, J. W. (1998). *Qualitative inquiry and research design: Choosing among five traditions.* Thousand Oaks, London, New Delhi: Sage.

Fielding, N. (2007). Computer applications in qualitative research. In P. Atkinson, A. Coffey, S. Delamont, J. Lofland, L. Lofland (Eds.), *Handbook of ethnography* (pp. 453–467). Los Angeles, London, New Delhi, Singapore: Sage.

Fielding, N. G., & Lee, R. M. (Eds.). (1991). *Using computers in qualitative research.* London, Newbury Park, New Delhi: Sage.

Fielding, N. G., & Lee, R. M. (Eds.). (1998). *Computer assisted qualitative research.* Newbury Park: Sage.

Glaser, B. G. (1978). *Theoretical sensitivity.* San Francisco: The Sociology Press.

Kelle, U. (Ed.). (1995). *Computer-aided qualitative data analysis.* London: Sage.

Kelle, U. (2005). Komputer-assisted qualitative data analysis. In C. Seale, G. Gobo, J. Gubrium, D. Silverman (Eds.), *Qualitative research practise* (pp. 473–489). London, Thousand Oaks, New Delhi: Sage.

Lewins, A., & Silver, Ch. (2004). *Choosing CAQDAS software: CAQDAS networking project.* Guildford: University of Surrey.

Saillard, E. K. (2011). Systematic versus interpretive analysis with Two CAQDAS packages: NVivo and MAXQDA. *Forum: Qualitative Social Research, 12*(1). http://nbn-resolving.de/urn:nbn:de:0114-fqs1101345. Accessed 17 February 2015.

Schönfelder, W. (2011). CAQDAS and qualitative syllogism logic—NVivo 8 and MAXQDA 10 compare. *Forum: Qualitative Social Research, 12*(1). http://nbn-resolving.de/urn:nbn:de:0114-fqs1101218. Accessed 17 February 2015.

Strauss, A. L., & Corbin, J. (1990). *Basics of qualitative research.* New Delhi: Sage.

Strauss, A. L., & Glaser, B. G. (1967). *The discovery of grounded theory: Strategies for qualitative research.* Chicago: Aldine Publishing Company.

Wiltshier, F. (2011). Researching with NVivo. *Forum: Qualitative Social Research, 12*(1). http://nbn-resolving.de/urn:nbn:de:0114-fqs1101234. Accessed 17 February 2015.

Unpacking the Relationship Between Learning to Read and Mental Health: Using an Ethnographic Case Study Approach

Jenn de Lugt and Nancy Hutchinson

Abstract Many children and youth who struggle to read experience the effects in both their school lives and their personal lives. Learning to read can and often does have benefits that extend well beyond the classroom. This multi-method, multi-perspective qualitative study follows three boys as they transition from struggling readers to more competent readers while participating in an intensive reading intervention. The purpose of this study is to further our understanding of the relationship between improved ability to read, and mental health. Dynamic stories are told about the three boys through ethnographic observations, and formal and informal interviews, with data provided by the reading instructors, classroom teachers, parents, the researcher, and the students themselves. The findings clearly show that learning to read has more than academic merit with a discernible impact on the mental health of the three boys who had once struggled to read.

Keywords Mental health · Reading · Behaviour · Emotional and behavioural disorders (EBD) · Self-efficacy

1 Introduction

Reading is arguably the most important academic skill for children to accomplish, often considered the cornerstone to all learning. This paper reports the findings of a qualitative study that describes the process of learning to read, and the concomitant changes in the mental health of three students. Despite experience and conventional wisdom acknowledging the interrelatedness of reading difficulties and poor mental health or problem behaviours, there has been surprisingly little work in this area.

J. de Lugt (✉)
Faculty of Education, University of Regina, Regina, Canada
e-mail: jenn.delugt@uregina.ca

N. Hutchinson
Faculty of Education, Queen's University, Kingston, Canada
e-mail: hutchinn@queensu.ca

© Springer International Publishing Switzerland 2017   95
A.P. Costa et al. (eds.), *Computer Supported Qualitative Research*,
Studies in Systems, Decision and Control 71,
DOI 10.1007/978-3-319-43271-7_9

This research uses a multi-perspective, ethnographic case study approach to show the evolution of three boys, in an intensive reading intervention, as they become more competent readers. Data were provided by the students, their parents, their classroom teachers, two reading instructors, and the researcher.

1.1 Study Purpose

The purpose of this research was to report descriptions of the lived experiences of three boys with both reading and mental health challenges as they progressed through an intensive reading intervention. Through a synthesis of the data I hope to enhance our understanding of how improving the ability to read for struggling readers may simultaneously benefit their mental health. The guiding research questions then are:

(1) In what ways does an intensive reading intervention affect struggling readers?
(2) How are changes in reading accompanied by changes in the mental health of previously struggling readers?

1.2 Context

The reading intervention, offered in a mid-sized city in central Canada, was research-based and used direct instruction to provide systematic and sequential lessons that focused on the development of reading decoding skills and, to a lesser extent and, when necessary, comprehension. This intensive intervention program, with one-on-one instruction, remediated reading difficulties of students who were reading at least one year, and usually two years, behind their peers. The instructors taught foundational and developmental reading skills including phonemic awareness, phonics, fluency, vocabulary, and comprehension; skills that are consistent with those identified by the National Reading Panel (NRP) as essential components of an effective reading program (National Institute of Child Health and Human Development 2000). They also included morphology, based on the findings of recent studies (e.g., Carlisle 2010). The reading intervention was structured and systematic while remaining sufficiently flexible to meet the needs of individual students. The Reading Room, a pseudonym, was privately operated, which limited availability to students whose parents could afford the fees or could obtain a bursary from a community agency or organization. The students in the program had generally experienced substantial and sustained difficulties in reading prior to enrolling in the program.

2 Rationale

This study sought to unravel and comprehend the complex and intertwined relationship between improved ability to read and mental health. There have been many studies with varied findings about this relationship. Research has shown that poor mental health adversely affects achievement (e.g., Murphy et al. 2015), and vice versa, poor achievement negatively affects mental health (e.g., Morgan et al. 2012). This study explored the ways in which mental health changed as struggling readers learned to read.

There has been a paradigm shift in the literature as reading difficulties in the past were shown to be associated with problem behaviours or emotional and behavioural disorders (EBD); it is now recognized that behavioural concerns are often manifestations of underlying mental health concerns. The earlier research reported that students with EBD often had coexisting academic and reading deficits (Levy and Chard 2001; Strong et al. 2004). Studies have also shown that problem behaviours (Trout et al. 2003), attention (Barriga et al. 2002), and more recently mental health (e.g., Dahle et al. 2011) are related to reading achievement. Specifically related to mental health, students with reading difficulties have been found to be characterized by feelings of low or inadequate self-efficacy (Ingesson 2007), feelings of anxiety (Dahle et al. 2011; Grills-Taquechel et al. 2012), somatic complaints, depression, and in some cases suicidal ideation (e.g., Dahle et al. 2011).

3 Methodology

This qualitative study was designed to create a detailed descriptive account of the experience of each of three boys as they progressed through the intervention, including any changes noted in their academic achievement, reading attitude, general and reading self-efficacy, attention, and their mental health. Reading ability was measured pre- and post-intervention using the *Test of Word Reading Efficiency* (TOWRE), subtests of the *Woodcock-Johnson*, and the *Gray Oral Reading Test-Fourth Edition* (GORT-4). Progress throughout the intervention was gauged qualitatively by observing what was being taught (word level difficulty and passage difficulty), the rate at which students advanced (reading progressively longer words and more difficult passages), time on task, and changes in attitude towards reading (increased willingness to read and engagement in reading). Changes in social, cooperative, and interactive behaviour were also recorded through observations, informal interviews with the students and the reading instructors throughout the intervention, and through formal semi-structured interviews with all participants following the intervention.

3.1 Participants

Three boys, Dillon, Logan, and Mason, ranging from Grades two to four began an intensive one-on-one reading intervention because of sustained difficulties they had experienced learning to read. Dillon came from a disadvantaged and challenging background, and his foster mother feared that he would not learn to read, because at age seven he did not know the sounds of the letters. He began the intervention as a non-reader when he was in Grade two. In addition, Dillon had been identified with a behavioural exceptionality. Although Logan was a struggling Grade 3 reader, his story unfolded in stark contrast to Dillon's. Logan came from a book-rich, supportive, and stable family, and was identified as having both a learning disability and as being extremely gifted (at the 99.9 percentile on an intelligence test). Logan struggled with decoding, but was still able to comprehend, and began the intervention with his sense of self reasonably intact. His mother described some emotional and behavioural concerns she had noticed with Logan, but his classroom teachers hadn't noticed any. Mason had a similar upbringing to Logan, but he began the intervention clearly affected by too few successes in school over too many years. Mason was able to decode reasonably well, but lacked fluency and comprehension. As the oldest of the three boys, he began the program when he was in Grade four—by this time he was becoming aggressive at home, and had been diagnosed with depression. Through the intervention all three boys made gains in their reading ability and demonstrated additional gains that clearly extended well beyond the classroom.

3.2 Data Collection

In this multi-perspective study, data were collected from a number of sources. All classroom teachers responsible for teaching the students were invited to participate in this study; one was interviewed for each of Dillon and Mason. Because Logan was in French Immersion three classroom teachers were interviewed: his French language classroom teacher (Jalen) who also taught him social studies and visual art; his English language teacher (Sydney) who also taught him health; and Angie, his French Immersion teacher who taught him science, math, drama, dance, and physical education. Angie was also the Student Support Teacher at Logan's school and provided one-on-one or small group instruction for Logan. The reading instructors, Jenelle and Helen, alternated days of instruction for Logan; Allyson and Jenelle alternated days for Dillon and Mason. Data was also provided by the students themselves, as well as their mothers (Logan and Mason) and an aunt/foster mother (Dillon).

Data were collected through ethnographic observations, and formal and informal interviews. Each student participant was observed every other day that they participated in the reading intervention; Dillon was observed for approximately 40 h,

Logan approximately 24 h, and Mason approximately 35 h. Post-intervention formal, semi-structured interviews with parents, classroom teachers, reading instructors, and the students were conducted in order to gather information on their perspectives as to what had changed for the student since beginning the intervention. All formal interviews were audio-recorded and transcribed verbatim. Informal interviews with the reading instructors were conducted throughout the reading intervention of all three student participants, and were recorded by notes made at that time.

3.3 Data Analysis

This multi-method, multi-perspective study was designed to elicit as much information about each case as possible. The cases were analysed separately by incorporating data from all participant groups—the students, the parents, the reading instructors, classroom teachers, and the researcher. Through an analysis of this data, it was possible to describe the relationships among the students' growing ability to read and their learning, social behaviour, and mental health.

Standard methods of qualitative theme analysis (Patton 2002) were used to analyse the interview data, facilitated through the use of NVivo 10, a data management system that helped organize, manage, and analyse the data. Interview transcripts were coded in order to identify categories, themes, and patterns. An initial reading of the transcripts identified issues and ideas as they emerged (inductive approach), providing the groundwork for subsequent coding. The documents were then reviewed again in search of additional data that supported these initial coded categories (deductive approach). Categories were subsequently merged to form a more focused coding framework that led to the identification of emergent themes. Each case was initially analysed and described separately, recognizing its individuality and unique situation. Subsequently, however, a cross-case analysis was performed in order to discern and highlight differences and similarities among the three cases. The observation data was then used to augment the findings of the interviews by verifying and providing substantive evidence for the emergent themes. The findings reveal important connections between developing reading skills and the mental health of all three boys.

4 Results

To varying degrees, the three participants made gains in their ability to read. Logan made the greatest gains, while both Dillon and Mason made smaller gains than had been expected and, in Allyson's view, gains not consistent with the typical gains of students in the intervention. Of the three, Mason showed the least improvement with only small gains in fluency (rate and accuracy) and comprehension. Table 1

Table 1 Pre- and post-intervention reading comprehension scores

Student	Grade/Age (at initial assessment)	Number of sessions	Comprehension: Pre-intervention grade equivalent	Comprehension: Post-intervention grade equivalent
Dillon	2.2/8 year 8 month	93	<1.0	2.0
Logan	3.1/8 year 0 month	50	<1.0	4.2
Mason	4.4/9 year 5 month	72	2.2	2.4

provides a synopsis of some of the key information related to the reading of each of the three boys. Reading comprehension was assessed using the Gray Oral Reading Test (Revised-4) (GORT-4). Because the ultimate goal of reading is comprehension, it was used as a general indicator of reading improvement.

4.1   Case Study 1: Dillon

In spite of getting extra help at school and home and attending Kumon, Dillon continued to struggle with reading. When he started at The Reading Room, Dillon's reading was extremely limited. He did not recognize any sight words except his name which he identified by its first letter D, nor did he know the sounds of many letters (Tanya—his aunt/foster mother). Shelley, Dillon's classroom teacher, said that Dillon started the school year in September as a D student in reading and writing, but was "shining" in math, science, and social studies, as a B student.

In addition to his struggles with reading, Dillon experienced emotional and behavioural challenges. Shelley described Dillon as a "behaviour student" and was concerned for his emotional well-being. Jenelle considered Dillon to be a "fragile kid" who was "quite volatile" and although he was "very curious about things, and about the world, and learning" he was also guarded and "cautious in his responses."

Changes in Dillon: "When I started coming here … it made my life better." As Dillon's ability to read improved, so did his writing, his attitude towards reading, and his on-task behaviour. These elements might have been expected to contribute to improved learning in his school experience, but participants reported mixed observations and perspectives regarding his behaviour upon full-time return to school. However, all observations confirmed that while he was attending the reading intervention, he exhibited fewer inappropriate behaviours and developed enhanced self-efficacy for academic learning, particularly in reading and writing. He became happier and was less volatile and labile.

4.2 Case Study 2: Logan

All of his teachers reported that Logan was highly intelligent and that there was a clear discrepancy between his verbal ability and his reading and writing ability. The amazing aspect that teachers and reading instructors alike frequently made reference to was his uncanny ability to comprehend, despite his apparent difficulties with decoding (observation, October 27, 2010). Sydney described his reading: "it was so segmented … there was no flow to it, he was not a fluent reader at all even at that level. And yet … his comprehension was there, and in fact *beyond*, he could make inferences, and make connections." Logan was formally tested with the results indicating that he was gifted with a learning disability.

Stacie, Logan's mother, described Logan as "struggling in reading, sounding out little bits—it was painful." She perceived that Logan's reading difficulties led to frustration and problem behaviours prior to the intervention: "by the end of the day … he's just so frustrated; and the frustration and acting out, when you tried to get him to do stuff, he showed a lot of anger, and frustration." In contrast, his classroom teachers considered him to be a cooperative and compliant student who was "eager to please, pleasant, happy … questioning, very inquisitive, doesn't have a whole lot of social skills for the group type activities, he would rather be on his own by himself" (Angie). Sydney, who had known Logan since Kindergarten, described him as being "a very timid, shy, uncertain, cried easily, reluctant participant … who needed lots of encouragement and kindness and softness," and Jalen described him as being "quite quiet, and he likes to follow the rules, and he is not disruptive at all."

Changes in Logan: "my reading wasn't very good, and I really wanted to go, actually *really* wanted to go [to The Reading Room]." When Logan started to make gains in reading, participants noticed a series of changes they thought evolved along with his developing ability. Despite his challenges with reading Logan had managed to maintain his self-esteem and therefore no change in his self-esteem was reported by any of the participants. However, with changes reported in his academic learning in both English and French, a better attitude towards reading and school, improved behaviour at home, and improved emotional well-being as well as a sense of empowerment, learning to read was an important and powerful achievement for Logan.

4.3 Case Study 3: Mason

A psychoeducational assessment indicated that Mason had weaknesses in reading, writing, and processing speed and was identified as having a language-based learning disability. Mason was also diagnosed with a depressive disorder. According to Jenelle, Mason could decode single syllable words "adequately," but he did not have any multi-syllable word attack skills, and he lacked prosody. Jenelle described his reading as "fairly fatiguing, lots of errors, lots of misreads, substitutions and insertions, so really reading was not an enjoyable experience, or a fulfilling experience and I think it was quite frustrating." According to his classroom teacher, Sandra, writing was also effortful for Mason, describing it as "such a chore, and it's messy" she would often have to ask him to rewrite his work because it was illegible.

Sandra described Mason as "Just a great kid" who was not achieving his potential and was deliberately very quiet with a "please don't notice me" purpose. Similarly, Allyson considered Mason to be "very quiet, very obedient, always did what you asked, but … usually looked like a pretty sad, little kid," and Jenelle, described Mason as "well behaved, flat liner emotionally or so guarded that he lets very little out, helpful, cooperative, kind, interested, curious … and yeah, also shutdown."

Changes in Mason: "he just seems like a happier kid." For Mason, the change emphatically commented on in the interviews was the change in his emotional well-being—specifically his improved confidence and disposition. Although no-one specifically referred to his depression, most participants described him as a much happier person. Both his mother and classroom teacher described his general self-efficacy as extending beyond the classroom and into sports—he had become more confident and would now take the puck and even scored a few goals, something he would not even have attempted before.

5 Discussion and Conclusions

This study reports three cases, of three struggling readers, each reported by adults in their lives to be experiencing concomitant mental health concerns. In addressing the first research question—*In what ways does an intensive reading intervention affect struggling readers?*—this study provides evidence that an improved ability to read positively affected the achievement, attitude, ability to attend, behaviour, mental health, and sense of empowerment of the student participants. The second research question—*How are changes in reading accompanied by changes in the mental health of previously struggling readers?*—was also supported by the data in this study. Gains in the mental health of the student participants in this study are related to improved general self-efficacy (or confidence), self-efficacy for academic learning, and improved disposition with participants described as being "happier" and "smiling more."

It is not the purpose of this research to suggest causality but to instead illustrate the broad effects learning to read can have on the mental health of students. Not only do students benefit academically, but emotionally as well. Furthermore, as problem behaviours are often, at least in part, manifestations of challenged or poor mental health, behaviour problems are also likely to be mitigated when reading improves. Participants in this study reported both positive changes in social competence and a reduction in problem behaviours.

Improved mental health and behaviour clearly has positive repercussions for the individual student, their classroom, school, and home environments. Most telling are the reports by the two mothers that their sons' gains in reading also improved their family dynamics and home environments. By successfully teaching reading to struggling readers, who through time often struggle with respect to their self-concept and emotional well-being, we can also improve their mental health. We know how to teach reading; by doing so, students may potentially make gains in two important areas—in their mental health, as well as in their ability to read.

References

Barriga, A. Q., Doran, J. W., Newell, S. B., Morrison, E. M., Barbetti, V., & Robbins, B. D. (2002). Relationship between problem behaviors and academic achievement in adolescents: The unique role of attention problems. *Journal of Emotional and Behavioral Disorders, 10*, 233–240.

Carlisle, J. F. (2010). Review of research: Effects of instruction in morphological awareness on literacy achievement–an integrative review. *Reading Research Quarterly, 45*(4), 464–487. Retrieved from http://search.proquest.com/docview/762467527?accountid=13480

Dahle, E. A., Knivsberg, A., & Andreassen, A. (2011). Coexisting problem behaviour in severe dyslexia. *Journal of Research in Special Educational Needs, 11*, 162–170.

Grills-Taquechel, A. E., Fletcher, J. M., Vaughn, S. R., & Stuebing, K. K. (2012). Anxiety and reading difficulties in early elementary school: Evidence for unidirectional- or bi-directional relations? *Child Psychiatry and Human Development, 43*, 35–47.

Ingesson, G. S. (2007). Growing up with dyslexia: Interviews with teenagers and young adults. *School Psychology International, 28*, 574–591.

Levy, S., & Chard, D. J. (2001). Research on reading instruction for students with emotional and behavioral disorders. *International Journal of Disability, Development and Education, 48*, 429–444.

Morgan, P. L., Farkas, G., & Wu, Q. (2012). Do Poor Readers Feel Angry, Sad, and Unpopular? *Scientific Studies of Reading, 16*(4), 360–381. doi:10.1080/10888438.2011.570397.

Murphy, J., Guzmán, J., McCarthy, A., Squicciarini, A., George, M., Canenguez, K., & Jellinek, M. (2015). Mental health predicts better academic outcomes: A longitudinal study of elementary school students in Chile. *Child Psychiatry and Human Development, 46*(2), 245–256. doi:10.1007/s10578-014-0464-4.

National Institute of Child Health and Human Development. (2000). *Report of the National Reading Panel. Teaching children to read: An evidence-based assessment of the scientific research literature on reading and its implications for reading instruction: Reports of the subgroups* (NIH Publication No. 00-4754). Washington, DC: U.S. Government Printing Office.

Patton, M. Q. (2002). *Qualitative research and evaluation methods* (3rd ed.). Thousand Oaks, CA: Sage Publications.

Strong, A. C., Wehby, J. H., Falk, K. B., & Lane, K. L. (2004). The impact of a structured reading curriculum and repeated reading on the performance of junior high students with emotional and behavioural disorders. *School Psychology Review, 33*, 561–581.

Trout, A. L., Nordness, P. D., Pierce, C. D., & Epstein, M. H. (2003). Research on the academic status of children with emotional and behavioral disorders: A review of the literature from 1961 to 2000. *Journal of Emotional and Behavioral Disorders, 11*, 198–210.

Reverse Logistics Companies' Perspective: A Qualitative Analysis

Mélodine Gonçalves, Ângela Silva and Celina P. Leão

Abstract Reverse Logistics has been object of great interest essentially in the consumer's culture change, in the competitiveness and in the increased environmental, and obviously due to its economic potential. In Portugal, Reverse Logistics is an unfamiliar word in the business world, specifically in SMEs. The present research focus to describe the Portuguese companies' standpoint and knowledge based on two aspects: the concept and the return of products. To verify the different companies' perspectives, semi structured interviews were applied in ten Portuguese Companies, of different size and in diverse industrial sectors. The qualitative data analysis was developed with the support of the webQDA software. The interviews analysis gave the opportunity to understand that large companies are aware of Reverse Logistics, in contrast to the SMEs' scarce knowledge. Related to the Reverse Logistics strategies applied to the product returns, this research has showed that the reuse of the products or their sale to the scrap or recycling industries as the most common strategies, although some particularities depending on the type of industry.

Keywords Multicases study · Qualitative analysis · Reverse logistics · WebQDA

M. Gonçalves (✉) · Â. Silva
Faculdade Engenharia e Tecnologias, Universidade Lusíada, Largo Tinoco de Sousa,
4760-108 Vila Nova de Famalicão, Portugal
e-mail: melodine_05@hotmail.com

Â. Silva
e-mail: d1279@fam.ulusiada.pt

C.P. Leão
Centro ALGORITMI, School of Engineering, University of Minho, Campus de Azurém,
4804-533 Guimarães, Portugal
e-mail: cpl@dps.uminho.pt

© Springer International Publishing Switzerland 2017   105
A.P. Costa et al. (eds.), *Computer Supported Qualitative Research*,
Studies in Systems, Decision and Control 71,
DOI 10.1007/978-3-319-43271-7_10

1 Introduction

In today's modern business environment, Logistics evaluate not only direct flows but also allows companies to consider undirected flows activities and supply chain information as an important part of Logistics (Lopes 2009). The strong competitiveness, the short life cycle of products, the laws pressure and the ecological awareness are some examples that demonstrate the importance of the development of Reverse Logistics process (Lopes 2009; Pokharel and Mutha 2009).

In simple terms, Reverse Logistics can be thought for as the process of collecting and transporting used or unwanted products from a customer or retailer site to an appropriate facility where the remaining product value can be recovered. In other words, Reverse Logistics refers to the distribution activities involved in product returns, source reduction/conservation, recycling, substitution, reuse, disposal, refurbishment, repair and remanufacturing (Akdoğan and Coşkun 2012). Reverse logistics is rapidly becoming an integral component of retailers' and manufacturers' profitability and competitive position. Product returns are the most common aspect of reverse logistics. Still, most companies do not handle returns as important as they should since they do not consider as part of their core competencies. Increasingly, reverse logistics must be considered part of a successful growth strategy. Nowadays, is essential to have an asset recovery strategy. Returns, repairs, and used items can also have branding implications (Silva et al. 2013). Furthermore, an efficient Reverse Logistics system can transform an expensive and complex return process into a competitive advantage for the company. The benefits will be evident if the processes and the execution are well defined.

The present work is part of a wider research project that analyzes and characterizes Portuguese companies' perspective based on three aspects: The Concept, the Returns and the Environment. Rubio et al. (2008), Lambert et al. (2011), Reddy (2011) are some examples of relevant publications on Reverse Logistics with emphasis in concept, product returns and environmental impact. Herein, the investigation will focus on the companies' knowledge concerning Reverse Logistics Concept and on the Reverse Logistics strategies applied to product returns. The webQDA qualitative software (Neri de Souza et al. 2011) was used to analyze the contents of the interviews scripts helping understand the present companies' Reverse Logistics perspective.

2 Reverse Logistics

It is no surprise that almost every company is looking for ways to improve their name and image, increase sales, decrease costs and risks. But in such tough economic times, the easy cuts have been made and all of the simple process improvements have been put in place. It is here that Reverse Logistics comes. Many authors have suggested many definitions for this concept and the most widely accepted definition of Reverse Logistics (RL) comes from European Working Group on Reverse Logistics, REVLOG. They define it as: "the process of planning, implementing and controlling backward flows of raw materials, in process inventory, packaging and finished goods, from a manufacturing, distribution or use point, to a point of recovery or point of proper disposal always with the purpose of capturing value" (Rubio et al. 2008).

Effective RL focuses on the backward flow of materials from customer to supplier with the goals of maximizing value from the returned item and/or assuring its proper disposal. This may include product returns, source reduction, recycling, substitution and reuse of materials, waste disposal, refurbishing, repair and remanufacturing (Autry 2005).

Reverse logistics processes and research has traditionally emphasized green logistics, i.e., the use of environmentally conscious logistics strategies (Lambert et al. 2011; Autry 2005). While environmental aspects of RL are critically important, many firms are also recognizing the economic impact of RL (Klausner and Hendrickson 2000). As a result of the aforementioned pressure, companies have adopted environmental practices increasing the investment in clean technologies and in the redesigning of processes and organization (González-Torre and Adenso-Díaz 2006). RL contribute to the expansion of customer service, satisfying requirements and expectations. It will be a matter of time until to understand the importance of put the Reverse Logistics as a part of Logistics system and a prominent position in the companies (Lopes 2009; Pokharel and Mutha 2009).

Almost every day, a customer, for any number of reasons, returns a product. Maybe the product was defective, or not in accordance with the specified, or it was the wrong size, or they are unhappy with the functionality of the product (unfulfilled expectations), and sometimes customers return products because they discover an alternative product with better functionality after they have made the purchase and others reasons. Whatever the case up until recently, manufacturers spent relatively little effort addressing the causes and effects of product returns (Blanchard 2007).

Once a product enters the RL flow, the logistics manager has to decide where the product has to be sent: either returns to vender, to the landfill, or to the secondary market. There exist seven channels for disposing the products that have been returned to the manufacturer: the return to vendor, sell as new, sell via outlet or discount, sell to secondary market, donate to charity, remanufacture/refurbish and materials reclamation/recycling/landfill. Based on the condition of the returned product, contractual obligations with the vendor, and the demand for the product, the manufacturer has one or more of the above options to dispose the returned product (Reddy 2011).

A quick and efficient handling of returned product could also be critical in sustaining relationships and creating repeat purchases (Autry 2005). For this reason, firms are more willing than ever to accept returns from customers. As a result, liberal return policies have become a standard marketing practice and a major component of the corporate image for many firms in both business-to business and business-to-consumer markets. The complexity of managing damaged or defective goods, product recalls, maintenance and repairs, and recycling should make RL programs a high priority.

3   Research Method

In order to understand and analyze the different companies' perspective, the research group applied a qualitative methodology. Unlike what happens in quantitative research, in qualitative methodology the research is more focused to find the cause-effect relationship. In qualitative methodology, each true it is relates with a context and a specific time.

In this study, the events of interest are unique within a context of real life, featuring an exploratory character and aiming to answer the research questions. A case study was used to assess the strategic factors to clarify the questions that require further research (Schultmann et al. 2006). Case study means a close analysis of the practice, together with the circumstances and its characteristics leading to an understanding of the situation within its own context (Brito 2014).

The case was the knowledge and perception of the Portuguese companies regarding Reverse Logistics Concept and Strategies within the context, the companies' settings. Interviews, one of the most common methods of data collection used in qualitative research, has one of the most powerful ways to understand the others perspectives and it's a powerful tool to capture the diversity of descriptions and interpretations about what people know on the field.

3.1   Methodology

To fulfill the objectives of the present work, an exploratory and descriptive research in real context was applied based on a multiple case study (Yin 1994). The guidelines selected for the methodology were based on five steps: (1) determination and definition of the research questions; (2) selection of the companies in northern Portugal and definition of semi-structured interview guide; (3) preparation of the formal email of invitation to be sent, guide definition and rules establishment for confidentiality; (4) data collection procedure in the field according to company/researcher agreed schedule and data base case study development; (5) analysis of the qualitative data and report preparation.

3.2   Sample Data Characterization

The research explores companies' of different type of sector and sizes in northern Portugal. The study focused the North by the large number of companies/industry located in this region (INE 2014). Thirty companies were contacted by a formal email enlightening the project and requesting collaboration. Later, it became necessary to contact by telephone due to the low number of participations in this first stage. One of the reasons of this low acceptance in the first contact could be the lack of RL word knowledge. Sentences like: "*I don't know what it means, so I can't help you*", "*We don't practice it*", "*That is a modern name for Logistics?!*", were used as justifications and resistance to talk and to take part in this study.

Ten companies accepted to collaborate in this study. Apparently this sample size can be considered small, however, and according to the time limit restriction of the study, this value was considered to be adequate for the research offering support to research aims and objectives (Baker and Edwards 2012). In some way, the sample was balanced in terms of size and type of activity: four Large Enterprises (LEs) and six Small and Medium-Size Enterprises (SMEs). Concerning industrial sectors and type of activity, two companies from the automobile components sector, two in the cutlery industry, one in the food industry, one in the drink industry, three in the aluminum industry and one in the retail industry. The participants are managers (M) or logistics directors (LD). Eight participants have higher education (six in engineering and two in business management), and two have compulsory education. In turn, the years of service varies between 2 and 32 years (see Table 1 for more details). Only one of the participants was female.

Table 1 Sample Characterization

Enterprises	Size	Industrial sector	Role	Higher education	Years service/age/gender
1	SMEs	Cutlery	M	Compulsory education	32/58/male
2	SMEs	Cutlery	M	Business management	20/45/male
3	SMEs	Aluminum	M	Mechanical engineering	77/33/male
4	SMEs	Aluminum	LD	Industrial management engineering	2/27/male
5	SMEs	Aluminum	LD	Mechanical engineering	5/34/male
6	SMEs	Food	LD	Agronomic engineering	3/32/female
7	LEs	Automobile	LD	Compulsory education	22/45/male
8	LEs	Automobile	LD	Industrial management engineering	12/42/male
9	LEs	Drinks	LD	Chemical engineering	5/54/male
10	LEs	Retail	LD	Business management	7/35/male

LEs Large enterprises, *SMEs* small and medium-size enterprises, *M* managers, *LD* logistics directors

3.3   Data Collection

After the definition of the research questions (*R.Q.1* and *R.Q.2*, identified in Table 2), semi-structures interviews were preferred in order to collect data allowing an individual and comparative study, including, whenever possible, corroboration or opposing findings from literature reviews. Semi-structured interview were conducted since, nevertheless the questions pre-defined order it can easily be changed depending on the flow of the discussion, i.e., it is possible to address questions in the appropriate time according to the interviewee's answers.

Most of the questions considered for the semi-structured interviews were either taken directly or inspired by the methodology developed by Reddy (2011). It included 13 predetermined open-ended questions, divided into 3 main topics: (1) the concept, (2) the products returns and (3) the environment. In this study the two topics, the concept and the products returns, will be discussed. The interview guide was sent to all the ten participants that accepted to be part of the study, one week before the interview; due to the sensitiveness of the topic, this approach sought to enlighten the participants about the nature of questions that need to be answered, giving them time to reflect and thus to be prepared.

Each interview did not exceed one-hour length; all of them were scheduled in agreement with the participant and recorded, on audio format with the agreement of the interviewees and the anonymity and confidentiality of data were granted.

Table 2   Research and interview questions according to the research dimensions (questions)

Research questions	Interview questions	Research dimensions
R.Q.1: what is the Portuguese companies' perception on reverse logistics?	Have you heard about the term reverse logistics? If yes, what do you understand by it? In your point of view, what is the importance of RL? What are the principal reasons for returns? Nowadays, in your opinion, the customers are more demanding?	Concept
R.Q.2: what is the perspective of Portuguese companies concerning the products returns?	How are the returns process adopted by your company?	Product returns
	What activity(ies)/strategy(ies) you apply, in order to get value from your products returns?	
	What are the principal reasons for returns? According to Franco (2010) "companies aren't prepared for the returns." What is your opinion?	

4   Data Analysis

All participants' interviews were transcribed and validated by each one of the interviewee. Then, and after the approval, all the reading transcripts were translated into English. In this step, a special attention was taken into account in order to guarantee the contents meaning and its details. For the qualitative data analysis of the semi-structured interview (organization and systematization of data) the authors made use of the software webQDA. The webQDA software supports the analysis of qualitative data in a collaborative and distributed environment. It focused on researchers who work in multiple contexts and need to analyze qualitative data as an individual or in-group in a synchronous or asynchronous way. It offers online and real time collaborative work as well provides a service to support scientifically research. The structure and functional organization of webQDA software is divided into three parts: (1) Sources, (2) Coding and (3) Question. The first area the research can put the data in different type (image, text, video or audio). In the second area the research create the dimensions, categories, subcategories. Exist a connection between the sources and the coding that contribute to organize the data and establish structured and interconnected between information. The third area has a set of tools that will help the research to answer the data.

For a better understanding of interviews information, the dimensions considered were the two R.Q. itself divided into categories: Concept and Products Returns. Each category has specific subcategories: RL knowledge, RL definition, and RL importance as Concept subcategories; and returns reasons and strategies as Returns subcategories.

The process of coding was conducted in different phases: after reading the data as a whole, the pre-defined codes were identified and confirmed in each of the interview responses. During this process a focus attention on potential distinct subcategories in the data was also carefully considered. Table 3 summarizes the process: participant's answers partition into words, phrases or sentences; and compilation into different categories according to its context and number of occurrences.

4.1   Findings Discussion

The perspective of nine men and one woman with several years of experience contributed to understand how LEs and SMEs enterprises face the concept and products returns and to perceive possible differences between companies' sectors and size.

Based on the opinions and perceptions of these ten Managers/Logistics Directors it was possible to answer the *R.Q.1: What is the Portuguese Companies' perception on Reverse Logistics?*

Table 3 The questioning process and webQDA tools

Questions	webQDA	Category/sub-category
What kinds of companies know the term RL?	Search for frequent words	Knowledge of the term RL
Is there any relationship between the term knowledge and company size?	Data matrix	Knowledge of the term RL by company's size (LE or SME)
What has been the companies' response about the importance of RL?	Search for frequent words	Importance
What are the principal reasons for returns?	Search for frequent words	Returns reasons
Is there any relationship between reasons for the returns and industrial sector?	Data matrix	Returns reasons by Industrial sector
What activity(ies)/strategy(ies) are applied in order to get value from the products returns?	Search for frequent words	Returns reasons
Is there a relationship between the type of strategies and the type of size?	Data matrix	Product return strategies by company's size

The concept was focused in three subcategories as follows: the RL knowledge, definition and importance. The last two subcategories were analyzed only by the five companies who know the term.

The results suggested that all the LEs (four) are aware of RL, and by only one of the six SMEs. Note that the SME' participant that new the concept is the youngest one and has a degree on Industrial Management Engineering (Enterprise 4, Table 1). Having in mind the definition of reverse logistics by REVLOG (Rubio et al. 2008), three companies (two LEs and one SMEs) state identical meaning: "*It is everything since the consumption point to the origin*", "*it's the close of logistics circle*" and the others (Retail and Drinks, two LEs) defined it according to the importance that RL have in their specific company, for example: "*For us Reverse Logistics is an important stocks optimization tool.*" Each company gave a definition similar to the one that can be found in the literature. However, and according to Kivinen (2002), each person has a different perspective of RL concept because that depends on which area they are inserted. So, it is important that each person indicate how the RL needs to be understood by all parts.

Concerning the importance of RL, all the companies agreed the importance and the benefits of RL as help to the process return minimize costs and help the environmental issues (use recyclable raw materials, reuse and others).

Based on the sub-question *R.S.Q.1: What are the principal reasons for returns?*, productive reasons were mentioned by the most companies: "*There are many specific products that it's need a lot technique or visual attention… and sometimes it happen production mistakes that aren't detected or detected later.*". Followed by transportations faults: "*Poor packaging*", "*During the transport route, the product may be damaged...".* The aluminum and automotive companies indicated the same reason (production) and the other two big companies (drinks and retail industry) indicated different reasons (information flow and seasonal products), realizing that the reason connected with the type of industrial sector.

Concerning the second sub-question, *R.S.Q.2, What activity(ies)/strategy(ies) apply, in order to get value from your products returns?,* the strategies are connected to the direct recovery (Reuse and Outlet and Resale), disposal or reprocessing (Repair and Polishing), as illustrated in Fig. 1.

Concerning the RL strategies, it was observed that the elimination is the strategy most used by the companies (six SMEs and two LEs). Companies show the best way to make value from products returns is sell them to the scrap or industries that will transform the product in by-products. Moreover, the second strategy applied by seven companies (four SMEs and three LEs) is reuse. Although, this strategy was mentioned as a second option, six companies prefer reuse their products firstly and if they cannot be reused, they are eliminated. It was observed that all companies reuse their products except food company; two aluminum companies apply repair; one cutlery company transform their products by polish. The retail company was the only company to apply redistribution and resale products (seasonal products). It was interesting to understand that all Aluminum companies have the same opinion about use-recycled aluminum: *"It isn't easy to use recycled... by my knowledge the recycled aluminum hasn't a good behavior in some techniques, like anodizing, and the quality it's complete different if you use the first fusion."* Contrasting opinions were reported in Logozar et al. (2006), which refer that the properties of aluminum are not affected by the recycling and reuse.

Based on the two-sub questions (*R.S.Q.1* and *R.S.Q.2*) the second research question could now be analyzed: *R.Q.2. What is the perspective of Portuguese Companies concerning the products returns?*

All the ten Portuguese companies share the same opinion that the customers are more demanding, and the competiveness and the liberation returns politic are some factors associated (Lee and Lam 2012; Amini et al. 2005; Figueiredo 2014): *"Today the demand is higher because there is more competition"*; *"All the customers are demanding, some times is the price, other is quality and others is the binomial price-quality... so it's very hard!".*

It was interesting to observe that six companies (three SME's and three LEs) agreed on that companies are not prepared for returns following the thoughts presented by Franco (2010): *"We think in our business like a direct flow. We don't*

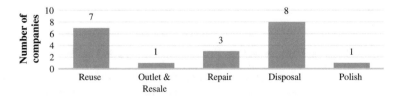

Fig. 1 Reverse logistics strategies related to the returns of products

think that what we produce someday can return", *"I agree for two reasons, the cost of packaging and the space that all those packs occupy."* On other hand, three SMEs and one LE disagree: *"Increasingly, the companies are more prepared because things need to change and people begin to have more attention to this type of situations".*

5 Final Remarks

This paper presents the Portuguese companies perspectives and knowledge on Reverse Logistics. For that, semi-structured interviews were conducted in ten Portuguese companies of different size (SMEs and LEs) and in different industrial sectors (Aluminum, Automotive, Drinks, Food, Cutlery and Retail), located in Northern Portugal. These ten companies voluntary accepted to be part of the research study. The analysis of the interviews gave the opportunity to understand and answer the addressed research questions regarding a new concept: Reverse Logistics.

All LEs are aware of Reverse Logistics, and just one of the SMEs knew the concept. The participant of this SME is the youngest one and has a degree on Industrial Management Engineering. This may be related to the different organizational cultural, competitiveness and financial support between large and SMEs. So, Reverse Logistics can be considered as new concept for the Portuguese SMEs.

Related to the Reverse Logistics strategies applied to the product returns, this research has showed that the most common strategies are the reuse of the products or their sale to the scrap or recycling industries, although there are some particularities depending on the type of industry.

Since this is an on-going work, more research in this area is essential. The research study continues not only with a in-depth analysis of the data collected but also by extending the study to other areas, namely continuing professional enterprises' strategies, and using different data sources.

Acknowledgments The authors would like to express their acknowledgments to national funds by COMPETE: POCI-01-0145-FEDER-007043 and FCT–Fundação para a Ciência e Tecnologia within the Project Scope: UID/CEC/00319/2013, and to all managers' companies who accepted the challenge to participate in this study.

References

Akdoğan, M. Ş., & Coşkun, A. (2012). Drivers of reverse logistics activities: an empirical investigation. *Procedia—Social and Behavioral Sciences, 58*, 1640–1649.
Amini, M. M., Retzlaff-Roberts, D., & Bienstock, C. C. (2005). Designing a reverse logistics operation for short cycle time repair services. *International Journal of Production Economics, 96*(3), 367–380.

Autry, C. W. (2005). Formalization of reverse logistics programs: A strategy for managing liberalized returns. *Industrial Marketing Management, 34*(7), 749–757.

Baker, S.E., Edwards, R. (2012). *How many qualitative interviews is enough?: Expert voices and early career reflections on sampling and cases in qualitative research.* Discussion Paper. NCRM (Unpublished). http://eprints.ncrm.ac.uk/2273/. Accessed 30 April 2016.

Blanchard, D. (2007). *Supply chains also work in reverse.* Industry Week. http://www.industryweek.com/articles/supply_chains_also_work_in_reverse_13947.aspx. Accessed 14 March 2016.

Brito, M. (2014). Managing reverse logistics or reversing logistics management? *Master thesis of series research in management,* University Rotterdam.

Figueiredo, P. (2014). Logística Inversa no Mercado de Telemóveis em Portugal. *Dissertação Mestre em Gestão de Serviços.* Faculdade de Economia da Universidade do Porto.

Franco, E. (2010). A Importância da Logística Reversa como um Diferencial Competitivo. Relatório Final de Curso em Logística Empresarial, Universidade Candido Mendes, RJ.

González-Torre, P. L., & Adenso-Díaz, B. (2006). Reverse logistics practices in the glass sector in Spain and Belgium. *International Business Review, 15*(5), 527–546.

INE. (2014). *Instituto Nacional de Estatística—Statistics Portugal.* https://www.ine.pt/xportal/. Accessed 21 May 2015.

Kivinen, P. (2002*). Value added logistical support service: outsourcing process of spare part logistics in metal industry, Part 2.* Research Report 138. Lappeenranta University of Technology.

Klausner, M., & Hendrickson, C. T. (2000). Reverse-logistics strategy for product take-back. *Interfaces, 30*(3), 156–165.

Lambert, S., Riopel, D., & Abdul-Kader, W. (2011). A reverse logistics decisions conceptual framework. *Computers & Industrial Engineering, 61*(3), 561–581.

Lee, C. K. M., & Lam, J. S. L. (2012). Managing reverse logistics to enhance sustainability of industrial marketing. *Industrial Marketing Management, 41*(4), 589–598.

Logožar, K., Radonjič, G., & Bastič, M. (2006). Incorporation of reverse logistics model into in-plant recycling process: A case of aluminium industry. *Resources, Conservation and Recycling, 49*(1), 49–67.

Lopes, D. (2009). Uma Contribuição na Estrutura dos Fluxos Logísticos Reversos das Lojas de Departamentos. *Master Thesis on Program in Transports Engineering,* Universidade Federal do Rio de Janeiro, Brasil.

Neri de Souza, F., Costa, A. P., & Moreira, A. (2011). Questionamento no Processo de Análise de Dados Qualitativos com apoio do software WebQDA. *EduSer, 3*(1), 19–30.

Pokharel, S., & Mutha, A. (2009). Perspectives in reverse logistics: A review. *Resources, Conservation and Recycling, 53*(4), 175–182.

Reddy, D. (2011). A study on reverse logistics. *Master Thesis on Product and Process Development, Production & Logistics,* School of Innovation, Design & Engineering, India.

Rubio, S., Chamorro, A., & Miranda, F. J. (2008). Characteristics of the research on reverse logistics (1995–2005). *International Journal of Production Research, 46*(4), 1099–1120.

Schultmann, F., Zumkeller, M., & Rentz, O. (2006). Modeling reverse logistic tasks within closed-loop supply chains: An example from the automotive industry. *European Journal of Operational Research, 171*(3), 1033–1050.

Silva, D. A. L., Renó, G. W. S., Sevegnani, G., Sevegnani, T. B., & Truzzi, O. M. S. (2013). Comparison of disposable and returnable packaging: A case study of reverse logistics in Brazil. *Journal of Cleaner Production, 47,* 377–387.

Yin, R. (1994). *Case study research: Design and methods* (2nd ed.). Thousand Oaks, CA: Sage Publishing.

E-mentoring: A Content Analysis Design

Isabel Pereira, Rita Cadima, Hugo Menino and Inês Araújo

Abstract Through a project ERASMUS+, named CommonS, has been implemented a platform were it will be possible to create communities of practice and e-mentoring to help on the development of soft skills. This tool could help higher education institutions to respond to the need of recruiters and companies on these skills development and prepare students to be working ready graduates. In this paper it is operationalized a data analysis type, specifically a design of content analysis, to implement with the data that will be collected inside each e-mentoring project. Being e-mentoring a new phenomenon to the people involved, our aim is to understand how it works and identify the best practices.

Keywords Content analysis · E-mentoring · Soft skills · Employability

1 Introduction

Internationally some companies are withdrawing their demands of a degree on recruiting process, justifying that having a degree does not mean that applicants have the necessary skills (Burke 2016). This reality should be a warning to higher education institutions when planning and evaluating their training offerings, without prejudice to its scientific and pedagogical autonomy, they should be concerned about finding new strategies to the actual needs of companies.

I. Pereira · R. Cadima · H. Menino · I. Araújo (✉)
Polytechnic of Leiria, Leiria, Portugal
e-mail: ines.araujo@ipleiria.pt

I. Pereira
e-mail: isabel.pereira@ipleiria.pt

R. Cadima
e-mail: rita.cadima@ipleiria.pt

H. Menino
e-mail: hugo.menino@ipleiria.pt

© Springer International Publishing Switzerland 2017
A.P. Costa et al. (eds.), *Computer Supported Qualitative Research*,
Studies in Systems, Decision and Control 71,
DOI 10.1007/978-3-319-43271-7_11

117

At the center of this question are the soft skills, since the technical and scientific expertise or hard skills are not the concerned issue. Soft skills are defined as "an umbrella term covering various survival skills such as communication and interpersonal skills, emotional intelligence, leadership qualities, team skills, negotiation skills, time and stress management and business etiquettes" (Deepa and Seth 2013, p. 7).

Several methodologies have been implemented by various institutions all over the world from internships, formal and informal courses, workshops, serious games, mentoring (Vieira and Marques 2014; Wheeler et al. 2012; Pereira et al. 2016) to accomplish the development of skills that can improve employability.

However is important to find ways that can quickly adapt to the needs of students, recent graduates and professionals with different levels of expertise. Due to technological and market developments the need for skills will oscillate, being necessary to anticipate these changes in advance to prepare students and professionals (World Economic Forum 2016).

We intent with Commons project (co financed by the European Union Erasmus + program on Strategic Partnerships typology) to develop a tool that can help to satisfy the needs of students, recent graduates and professionals to improve their employability. In the CommonSpaces platform is already possible to create communities of practice, it can be formed by interests or specific needs of its members (Wenger 1998) where new groups may emerge as new skills are needed to be develop, adapting to the demands of job market, this is the major benefit of undertaking this project, the ease adaptation to the needs of members. Also, it will be possible to develop e-mentoring process between people that have the expertise and people who want help to learn it. With this functionality we expect that students could interact with future professional peers and develop skills needed to increase their employability odds.

With the beta version of the platform available, we started to form small groups to evaluate it. We will use the content analysis design presented bellow to identify the member's interests, difficulties felt on the platform, communication tools more used and the popular skills on e-mentoring type of projects. Our aim is to understand how an e-mentoring process works and if effectively helps members to improve the skills they need.

In this paper is our purpose to operationalize a data analysis type to perform with the data that will be collected inside each e-mentoring project. It will be presented below a design of a content analysis.

2 E-mentoring: A Content Analysis Design

Mentoring occurs when a person with more expertise guides and monitors the other with less experience or difficulty in achieving the sated objectives (Gay 1995).

This is a process that may occur formally when an institution promotes it, which implies that in addition to mentor and mentee exist a third party concerned that can impose limiting in terms of objective and time realization. Also this process can occur informally, because comes up from needs and interests of the mentor and mentee only, without having to answer to others. It is considered that promote informally the mentoring processes may result on an more intense experience for mentor and mentee, because are not imposed limits, allowing the progress to be centered on professional or personal needs, without schedules required by external factors (Janssen et al. 2015).

In a project developed by Wheeler et al. (2012) was implemented a process of e-mentoring enabling engineering students to be guided by professionals, thereby improving their employability potential. Through conversation tools (Skype, email and social networks) students could contact with professionals and understand what is expected from them on the recruitment, knowing how to improve their *curriculum vitae*. This project found that it is important to maintain regular contact between mentor and mentee. This experience allowed the students to connect with the professional world, preparing them for the recruitment phase.

In a review on mentoring in academic medicine is highlight that "Mentorship was reported to have an important influence on personal development, career guidance, career choice (…) but the evidence to support this perception is not strong" (Sambunjak et al. 2006, p. 1103). For this reason is recommended to perform studies using more rigorous methods. Is our intention to achieve this with our design that will be presented next.

2.1 *CommonSpaces E-mentoring Process*

CommonSpaces is a platform where is possible to communities of practice to grow around common interests, being the general thematic the skills that can help to improve employability. Each community can have subprojects that can be about specific topics, where is possible to catalogue Open Educational Resources (OER), or to build Learning paths (LP) with OER catalogued or develop an e-mentoring project.

To be a mentor, a member will have to follow some steps defined on the platform, but inside an e-mentoring project s/he has all the autonomy to guide the mentee. To be a mentee, each member has only to ask to be accepted by the mentor on the e-mentoring project.

On the platform are also available some communications tools, like: forums, internal mail, chat and video-chat. Also is possible to share files between the members inside each project.

2.2 *Methodology Design*

Our intention is to perform content analysis technique on the interactions made between mentor and mentee and identify the best practices. It will be implemented as an structural study, that is defined by Amado et al. (2013) as a way to highlight the regularity of a phenomena or its characteristics. Because this is new phenomena to the people involved, we want to document and understand what will happen, identifying the best practices which will help to create guidelines to help new members.

The members on the platform will be invited to begun e-mentoring projects with our assistance. They will be informed about our research and how the data will be collected. All exchange made inside, like OER, LP, files, and forum messages, will be register in a excel file to be analyzed afterwards with the help of software MaxQDA. For private messages and chats, we will ask to mentors and mentees respond a survey about it. All this data will become different documents inside of the MaxQDA software and will be categorized as described on Table 1.

This categorization of documents will be done by one coder and reviewed after by two other members of the project.

Is our intention to collect information, at least, from 15 e-mentoring projects, for reasons of validity. These categorization items are not close, because during the reading of the documents will be possible to find new items that could be important for our research.

At the end of each e-mentoring project we will do an interview with mentor and another with the mentee, so it will be possible to understand the benefits of e-mentoring and what need to be improved. It will be in this interview that we will collect information about the success of the e-mentoring experience, by asking how they evaluate their experience and if their goals were achieve at the end. The ones where both mentor and mentee have higher scores will be the successful ones.

Table 1 Documents to be collected and the categorization we intend to implement

Documents to collect	Description	Categorization to implement
OER catalogued	Each OER inserted into the e-mentoring project is cataloged according to predefined parameters	Who created (mentor/mentee) Skill involved (list) Type of OER (video, lecture,...)
LP created	Each LP inserted into the e-mentoring project is cataloged according to predefined parameters	Orientations made for LP (task, only explore, identify something, reflection, ...)
Documents shared	Each project have a sharable folder where is possible to upload different files. They will be cataloged according to predefined parameters	Who created (mentor/mentee) Skill involved (list) Type of document (lecture, task, example...)
Message exchange (forum, mail and chat)	The forum will be available to all members, but the chat and mail message no. It will be requested to mentor and mentee to send us a copy or a resume and to fill a survey about it	Who initiates (mentor/mentee) Skill involved (list) Type of message (question, opinion, clarification about an OER/LP, more information, orientation needs, task...)

3 Future Work

We expect that informal e-mentoring process can be a new way for students interact with the professional world during their graduation. Is something new, for these reason we want to collect all possible data to analyzed and understand how it will effectively work. With these information will be possible for us to create guidelines to new e-mentoring projects on the CommonSpaces, also will be possible to improve the platform to answer the needs identified.

To achieve this, we intend to perform the content analysis design presented in this paper to all data collected inside each e-mentoring project (minimum of 15). The results will help us to understand how informal e-mentoring happen, identify the best practices to guide new members and verify if effectively it improves employability.

Is the aim of CommonS project to provide a platform that could help the development of soft skills and answer to other needs felt by members to improve their employability. Informal e-mentoring could be an excellent choice, because make possible to students interact with professionals and learn directly from them (Janssen et al. 2015).

References

Amado, J., Costa, A. P., & Crusoé, N. (2013). A técnica da análise de conteúdo. In J. Amado (Ed.), *Manual de investigação qualitativa em educação* (pp. 301–351). Coimbra: Imprensa da Universidade de Coimbra.

Burke, L. (2016). Graduate jobs: University degrees no longer relevant to some employers. www. news.com.au. Accessed February 10, 2016.

Deepa, S., & Seth, M. (2013). Do soft skills matter?—Implications for educators based on recruiters' perspective. *IUP Journal of Soft Skills, 7*(1), 7–20.

Gay, B. (1995). What is mentoring? *Education + Training, 36*(5), 4–7.

Janssen, S., van Vuuren, M., & de Jong, M. D. T. (2015). Informal mentoring at work: A review and suggestions for future research. *International Journal of Management Reviews*, p. n/a–n/a.

Pereira, I., et al. (2016). Desenvolver competências que melhorem a empregabilidade. In *Tic Educa 2016*. Lisboa: Instituto da Educação da Universidade de Lisboa (in press).

Sambunjak, D., Straus, S. E., & Marušić, A. (2006). Mentoring in academic medicine: a systematic review. *JAMA, 296*(9), 1103–1115.

Vieira, D. A., & Marques, A. P. (2014). *Preparados para trabalhar?*. Porto: Um Estudo com Diplomados do Ensino Superior e Empregadores.

Wenger, E. (1998). *Communities of Practice: Learning, Meaning, and Identity*. In R. Pea, J. S. Brown, & J. Hawkins (Eds.), Cambridge: Cambridge University Press.

Wheeler, A., Austin, S., & Glass, J. (2012). E-mentoring for employability. In Centre for Engineering and Design Education (Ed.), *EE2012—Innovation, Practice and Research in Engineering Education*. (pp. 1–9). Loughborough University .

World Economic Forum. (2016). *The Future of Jobs, Geneva*. Available at: http://www3.weforum. org/docs/WEF_Future_of_Jobs.pdf. Accessed January 25, 2016.

Occupational Therapy's Intervention on Mental Health: Perception of Clients and Occupational Therapists About Intervention Priorities

Jaime Ribeiro, Pedro Bargão Rodrigues, Ana Marques, Andreia Firmino and Sandra Lavos

Abstract Literature points out that not always the intervention priorities considered by clients meet with those outlined by Occupational Therapists (OT). The intervention of OT in Mental Health aims to understand how occupations are amended in accordance with client expectations. The research here described seeks to understand the perception of clients and OT on intervention priorities, trying to determine whether there is compliance between the views. Within a qualitative approach was carried out a descriptive and exploratory case study with data triangulation from different sources. Software aided content analysis technique was used for the interpretation of data obtained through semi-structured interviews. OT and clients have differing views regarding intervention and its priorities. While the OT prioritize habits and routines, especially related to the ADL and IADL, customers give special focus to problem solving. It was found that clients are not fully satisfied for not being allowed to work for individual goals.

Keywords Occupational therapy · Mental health · Intervention priorities · Clients · Occupational therapist

J. Ribeiro (✉) · P.B. Rodrigues · A. Marques · A. Firmino · S. Lavos
School of Health Sciences, Polytechnic Institute of Leiria, Leiria, Portugal
e-mail: jaimeribeiro@ua.pt

J. Ribeiro
Health Research Unit (UIS), Polytechnic Institute of Leiria, Leiria, Portugal

J. Ribeiro
Inclusion and Accessibility in Action Research Unit (IACT),
Polytechnic Institute of Leiria, Leiria, Portugal

J. Ribeiro
Research Centre Didactics and Technology in Teacher Education—CIDTFF,
University of Aveiro, Aveiro, Portugal

© Springer International Publishing Switzerland 2017   123
A.P. Costa et al. (eds.), *Computer Supported Qualitative Research*,
Studies in Systems, Decision and Control 71,
DOI 10.1007/978-3-319-43271-7_12

1   Introduction

When Occupational Therapists intervene in mental health, they seek to understand how occupations are influenced by the client's life expectations and specific events.

Occupational Therapists help people perform the activities they need through the therapeutic use of occupations and outlining their intervention based on what is important to the client. The Occupational Therapy intervention usually includes an individual assessment in which, together with the client/family, the objectives of the intervention are delineated. A custom made intervention is structured to improve the person's ability to perform daily activities and reach goals, as well as a result assessment to ensure that the goals are being met and/or change the intervention plan to be able to meet them (AOTA 2016).

One of the professional models usually used by Occupational Therapists is the Model of Human Occupation, which conceptualizes people's characteristics in three interrelated elements: volition, habituation and the ability to perform (Kielhofner 2009). Volition is related to a person's motivation for the occupation and is related to life experiences. As volition is conditioned by life experiences, it conditions the priorities identified for his/her rehabilitation process. Habituation refers to the way that the person organizes his/her performance in routines and roles. Finally, the ability to perform refers to the client's skills to perform his/her activities. Therefore, the continuity of the intervention is essential to understand the coping abilities in what refers to changes in life (Creek and Lougher 2008).

In a study conducted in Mexico City in 2011 by Rivas-Garibay et al. (2011) for the purpose of identifying the benefits that clients believe they have obtained from Occupational Therapy, it was concluded that they identify the following main benefits of Occupational Therapy for their lives: the improvement of their inter-personal and family relations, the acquisition of new skills, the increase in the feeling of well-being and self-esteem and the fact that they get some economic benefits. However, and although most results state that Occupational Therapy fulfills its goals, the study mentions that it is important to implement new measures to help provide better care and treatment quality for psychiatric clients.

Another study conducted in South Africa in 2014 by Smith et al. (2014) iden-tified and compared the intervention priorities of the Occupational Therapists and of the clients with a schizophrenia diagnosis. It concluded that both clients and Occupational Therapists perceive family, social and friend's support as a priority in order to avoid readmissions. However, these perceptions are different in most performance factors, showing how important it is for Therapists and clients to establish the same therapeutic goals.

There are similarities in both studies regarding what the clients and Occupational Therapists consider a priority in the intervention, but there are frequent differences in the expectations of clients and professionals.

The background for this study emerged of a detailed bibliographical revision on how mental illness clients perceive their participation in Occupational Therapy, through which it has been concluded that there are some similar studies conducted in other countries.

Therefore, it was considered pertinent to make a case study in Portugal in order to compare its results with the remaining studies that have already been performed because there is no investigation that approaches this issue in what concerns people in Portugal. Thus, our initial problem is based on knowing how clients and occupational therapists perceive the intervention priorities in a mental health context.

2 Methodology

The present study has a qualitative approach, i.e. a study in which the researcher is the key tool and collects data in the natural environment, and this translates into a case study with a descriptive-exploratory objective. In this case, data will be collected in the natural environment of the clients and Occupational Therapists of the Hospital Infante D. Pedro—Aveiro, allowing for a more comprehensive view of the issue.

Three Occupational Therapists and three clients of the Department of Psychiatry and Mental Health of the Hospital Infante D. Pedro—Aveiro were interviewed. The equitable choice of the six participants is justified by the fact that there are three Occupational Therapists working at the above-mentioned service and this way the triangulation of the data collected by these methods is facilitated. The professionals working at the service initially proposed a group of four clients to participate in the study, from which one was excluded because he showed signs of dementia. The participants were chosen randomly, fulfilling the following criteria: male and female clients regardless of age with any mental problem that are committed to the Department of Psychiatry and Mental Health of the Hospital Infante D. Pedro— Aveiro; clients that accept to be interviewed and that signed the Informed Consent Form; clients that can read and write; Occupational Therapists with a minimum work experience of 3 years in Occupational Therapy and of 6 months in the Department of Psychiatry and Mental Health.

Considering that the interests of the clients are one of the study's main points, the application of the Occupational Self-Assessment (OSA), which is an instrument used to assess the above-mentioned model, seemed pertinent. This instrument was applied because it allows us to know some of the clients' interests. Besides the application of OSA, the clinical process of the clients involved in the study was also analyzed (Sousa 2006).

An interview script was drawn up and validated by experts in order to make it credible and reliable. The validation was made by an Occupational Therapist with recognized experience in mental health and one Professor with Expertise in Research, respectively in what concerns content and apparent validity.

In order to examine the data, the content of the interviews was scrutinized trough content analysis subsequently to a digital verbatim transcription. The content analysis consists of a set of research techniques that aim at finding the sense of the information collected through the interviews recorded on a digital recording and later transcribed (Campos 2004). This data analysis technique was implemented in accordance with the premises of Bardin (2015) and concretized with the use of dedicated qualitative data analysis software.

The software webQDA was used to assist in the analysis of qualitative data and allows for the analysis to be made individually or in collaboration (Souza et al. 2011). The different answers obtained in the interviews were coded in this software according to the rule of exclusivity.

After importing of the corpus of analysis into WEBQDA the sources were classified and attributes labeled. Posteriorly, data was coded into hierarchical trees of predetermined (identified during the interview guide creation) and emerged categories.

Computer-assisted qualitative data analysis (CAQDAS) allowed data manipulation and systematization as well enabled keeping track of notes, data sorting and rearrangement and a useful quick revision of coding decisions.

Pragmatically, CAQDAS permitted easier, flexible and accurate data analysis in each category, also allowed to assess quantity of registration units allocated to each category and the percentage of respondents' speech integrated in each of the categories subject to analysis. The category frequency analysis allowed to identify what might be considered by the individuals as relevant aspects in their rehabilitation process.

Counting of occurrences (explicit and implicit) per category is assumed by several authors as a way to determine major apprehensions of the interviewed, as words/ideas/thoughts repeated most often are the ones that reflect paramount concerns. Although quantifying speech can be a starting point, it can't be a binding for inferences about matters of importance (Stemler 2001).

The triangulation was potentiated trough a more searchable, viewable and manipulable data from the different sources used.

From now on, the term client(s) will be used instead of patient(s)/user(because it is the most correct term to be used according to the terminology of the 2nd Version of the Occupational Therapy Practice Framework (AOTA 2014).

3 Ethical Aspects

Regarding ethical aspects, it is important that at the start of each interview all participants have signed an Informed Consent Form, in which it is guaranteed that the data is only used for scientific promotion purposes. It is guaranteed that the identity of the participants is not disclosed by the use of the initials "OT" for the Occupational Therapists and "CL" for the clients followed by 1, 2 and 3 in order to distinguish them from each other.

The collection of data only started after permission from the Ethics Commission of the Hospital that participates in the study.

4 Data Analysis and Discussion

The subsequent discussion is organized by category used in the content analysis. Thus, in the text below arise dissected the categories used for encoding and interpretation of its contents.

4.1 Barriers in the Process of Rehabilitation

The first defined category concerns "Client Barriers—Perspective of the Occupational Therapist", through which it can be concluded that family is the client's main limitation when admitted to the service. According to OT2 and OT1, family support is very limited, either because family members are old or because they don't understand the pathology. From OT3's interview, one learns that this eventually influences the intervention of Occupational Therapy, because therapists are not always able to have access to the family when they need it. This is consistent with previous studies that also highlights that clients and therapists agree that the fact that has greater impact on the readmission is the support (both family, friends or community) (Smith et al. 2014). Besides family, therapists say the main limitations are low education levels, financial difficulties, substance use, medication, routines and the motivation of clients. However, Occupational Therapists identify some facilitators of the clients' occupational performance, like the support given to Occupational Therapy activities and motor skills. Being committed again was also mentioned by TO2 as a strong point for the following reason: *"Those that have been committed more than once are more open to giving us information about their homes and themselves, more receptive to new interventions and activities because they already know how things work. It is easier to create a therapeutic relationship"*.

4.2 Intervention Priorities

The individual's occupational performance is influenced by several factors and it is important that the several occupation areas are balanced so that the individual is functional. This being said, there are aspects that should be worked on first in order for that functionality to be accomplished. In what concerns the category "Intervention Priorities—Perspective of the Occupational Therapists", which includes the units that are part of the identification by the Occupational Therapist of the intervention priorities for their clients, it can be concluded that great importance is given to Activities of Daily Living (ADL), equitatively followed by productive and leisure activities and, finally, Instrumental Activities of Daily Living (IADL) *"The ADL, because they end up neglecting a lot and (...) they do it without giving it much meaning and we try to explain to them that importance"*, mentions OT2;

"They bring that part (ADL) when they are really down, meaning they neglect it completely", says OT1. Besides that, OT2 states: *"Leisure activities also end up being activities neglected by the clients"*, which is confirmed by OT1,

> About leisure time, they don't know how to spend the free time we have every day; or they are maniac and they spend the day doing anything but something objective; or the depressive and psychotic do practically nothing and they stagnate, they really stop, stop thinking, stop thought, stop everything.

As was mentioned before, Occupational Therapists recognize the importance of working in productive activities. OT3 mentions: "In the productive activities for them, paid or volunteer work, something they can do, even if it is not paid, but that makes them have a routine". OT3 also recognizes the importance of implementing routines for the clients. On their hand, clients only give priority to solving problems. CL3 says: "I hope to be prepared for life in the future"; and CL2 explains: "I hope to be able to better overcome the obstacles in my life; to acquire techniques that help me overcome my problems". It is therefore concluded that, in what concerns intervention priorities, the opinions of Occupational Therapists and clients diverge, just as stated in an international study. The study mentions that perceptions differ in most performance factors and solves this problem highlighting how important it is that therapists and clients establish the same therapeutic goals (Smith et al. 2014).

4.3   The Interests of Clients and Their Importance for Planning the Intervention

The intervention of Occupational Therapy planned according to the interests of the clients facilitates its success because it increases the motivation of clients and promotes an active participation in the activities. This opinion is based on the Human Occupation Model, which mentions that Therapists should assess and understand the client in order to develop, implement and monitor an intervention plan that considers their needs, interests and concerns (Kielhofner 2009).

Regarding the category "Interests", which focuses on the clients' perception of activities they are interested in performing, one can observe that there are many answers in the area of leisure, which is mentioned by all three clients as being important for them. The area of education was also mentioned by CL2 as follows: *"I like to study languages and literature, to read a lot"*.

It is well-known that the Occupational Therapist has to consider the opinion and the interests of the client in order to promote a greater interest in, and satisfaction with, the treatment. OT3 confirms this: *"The opinion of the client is one of the most valuable things and we as Occupational Therapists have to consider that, and there is always a negotiation between the client and the Occupational Therapist."* OT3 further states: *"The interests of the client influence my intervention because I have to set the goals according to the interests of the client and not my own"*. However, although the opinion of the three Occupational Therapists is unanimous, OT2 recognizes: *"The way the facilities work doesn't always help us to undertake those meaningful activities because we work with heterogeneous groups (...), but we always try to consider it when it is possible".* It can be concluded that although clients identify and communicate their interests to Occupational Therapists and these professionals recognize the importance of these interests for the success of their intervention, the way the service works doesn't make it possible to work towards implementing the activities that are meaningful for each client. Due to the number of recipients of Occupational Therapy treatments, there are group sessions in order to work with as many clients as possible.

4.4 Assessment and Satisfaction

The impossibility of working towards individual goals in Occupational Therapy at the Hospital Infante D. Pedro—Aveiro is recognized by the clients in the category "Client Satisfaction". *"Sometimes, not everything is as we would like it to be"*, says CL3. CL2 confirms: *"We never talked about what I wanted to do. It was more a choice for the whole group, not something individual".* However, although they work for a group goal and recognize that their interests are not always considered, the clients' assessment of the intervention of Occupational Therapy is positive. Just like CL2 says in the category "Appreciation of the Intervention—Perspective of the Client": *"Occupational Therapy is important because it gives me the tools that will support me more when I go outside. To have at least the bases to know how to behave, how to handle things in a difficult situation (...)."* The same is mentioned in an international study that says that clients think they have benefited from the acquisition of new skills in Occupational Therapy (Rivas-Garibay et al. 2011). This assessment is confirmed by the active participation in the activities in order to reach the goals and through verbalization, by telling their family members they like it and highlighting the importance of Occupational Therapy for themselves, just like OT2 and OT3 said in the category "Assessment of the Intervention—Perspective of the Occupational Therapist".

However, the Occupational Therapists were also dissatisfied, in the category "Occupational Therapists Satisfaction", with the results of their intervention and with the impossibility of working on individual goals with the clients. OT1 says: *"We have heterogeneous groups and it is often also difficult to work with them individually".* While OT2 says that: *"It is inglorious. The client is here for 16 days and never comes on the 1st day. Sometimes he/she only comes after a week. We are with him/her for a week and then he/she leaves. Clients are discharged without our knowledge. This isn't even discussed as a team. The doctor discharges them and that's it. Sometimes it is the client who warns us that he/she is leaving and that ends up not helping us as an OT because we are not able to finish the process since the client is lost. Working only here at the hospital and not being able to go the person's home is also a huge limitation."*

Table 1 Results obtained with OSA

	Step 1: degree of ease with which you perform the task			Step 2: importance of the activity			Step 3: 4 items you would like to change in you		
	CL1	CL2	CL3	CL1	CL2	CL3	CL1	CL2	CL3
Focusing on my tasks	Very well	Very well	Well	Very important	Quite important	Quite important			x
Being physically able to perform tasks that I need	Well	Very well	Very well	Very important	Quite important	Very important	x		
Taking care of the place I live in	Many problems	Well	Some difficulty	Important	Not very important	Important	x		x
Taking care of me	Very well	Very well	Well	Very important	Very important	Very important			
Taking care of others for whom I am responsible	Very well		Well	Very important		Very important			
Being able to go where I need to go	Very well	Well	Some difficulty	Very important	Important	Quite important			
Manage my money	Well	Well	Well	Very important	Quite important	Quite important		x	x
Manage my basic needs (food, medicine)	Well	Well	Well	Very important	Important	Very important	x		
Being able to communicate with other people	Very well	Very well	Well	Important	Quite important	Quite important			
Getting along with other people	Well	Very well	Well	Important	Quite important	Quite important			
Identifying and solving problems	Well	Some difficulty	Some difficulty	Very important	Important	Quite important	x	x	x

It would be beneficial if the decision of discharging the clients would be made as a team so that the Occupational Therapist would be able to better adapt his/her intervention.

4.5 Occupational Self-assessment

Occupational Self-Assessment (OSA) is a standardized assessment scale that reflects the individual character of the results and the needs of each client, which makes a client-based intervention easier. It was used in this study to understand the clients' perception about their skills using as means of comparison the information collected during the interviews (Table 1). Only the "about me" part, which was translated from Portuguese by Sousa (2006), will be used for this project.

It can be concluded that the clients do not identify IADL as an intervention priority because they mention how easily they perform most activities. However, CL1, CL2 and CL3 identify problem solving as a very important aspect for them, but one they find hard to perform, which makes them want to change it.

It is also important to highlight the fact that clients identify the ADL, namely "Taking care of me", as a "Very Important" task. However, unlike the Occupational Therapists, who consider that these activities are neglected in the clients' occupational performance, they say that they can do them "Very Well". It can therefore be concluded that clients have trouble identifying their limitations and/or ability to perform specific tasks.

It is concluded that there is a consensus between the information supplied during the interviews and the information collected through OSA.

5 Conclusions

After analysing the content, it can be concluded that there is a match between the collected data and the information obtained from the international studies mentioned above, namely in what concerns the benefits obtained from Occupational Therapy in Mental Health.

The Occupational Therapists and the clients that were interviewed show different opinions regarding the intervention of Occupational Therapy and its priorities. While Occupational Therapists give priority to the promotion of habits and routines, especially those associated with ADL and IADL, clients are particularly interested in problem solving.

In what concerns the satisfaction about the intervention of Occupational Therapy, it is safe to say that, according to the opinions that were presented, clients are not satisfied with the work that has been developed, although they give the therapists a positive assessment of the intervention. Though they identify some benefits of working in a group, they highlight the importance of working on individual goals. The professionals share the opinion, recognizing this omission, but they justify this impossibility of giving individual support with the service organization. It can therefore be concluded that it would be useful to adapt the service to the client and not the client to the service.

It is concluded that the main barrier to the client's occupational performance when he/she enters the service is, according to the Occupational Therapists, family support. Considering that the clients identify problem solving as one of their main limitations, it can be concluded that this limitation is related to the lack of family support, a barrier that was identified by the Therapists.

None of the clients was able to clearly identify any priority/interest and explain its association with their rehabilitation process. This means that during their interview they did not explain how it would be beneficial for their improvement to perform the activities identified as an interest. It would be pertinent to go deeper into this study to see if the clients' priorities/interests are suitable for the rehabilitation process.

The study can't be generalized, considering that is was only based on a small sample and in only one Hospital and Service of the country. Each Service and Occupational Therapist has different ways to organize and plan the intervention, as well as to value the interests and priorities of the clients.

It would therefore be important to conduct more studies focusing on the clients' and Occupational Therapists' perception of the intervention of Occupational Therapy in Mental Health, especially in the people of Portugal, so as to identify flaws and solve possible limitations.

Acknowledgments This paper is the result of a research work that involved students and professors of the Degree Course in Occupational Therapy—academic year 2014-2015. The authors acknowledge the collaboration of the Hospital Infante D. Pedro—Centro Hospitalar do Baixo Vouga, in particular the Occupational Therapists of the Psychiatry and Mental Health Service.

References

American Occupational Therapy Association (AOTA). (2014). Occupational therapy practice framework: Domain and process (3rd Edition). *American Journal of Occupational Therapy, 68* (Supplement_1), S1-S48. doi:10.5014/ajot.2014.682006

American Occupational Therapy Association (AOTA). (2016). About Occupational Therapy. Obtido de AOTA: http://www.aota.org/about-occupational-therapy.aspx

Bardin, L. (2015). Análise de Conteúdo. Lisboa: Edições 70.

Campos, C. J. (2004). Método de Análise de Conteúdo: ferramenta para a análise de dados qualitativos no campo da saúde. Revista Brasileira de Enfermagem, 611–614.

Creek, J., & Lougher, L. (2008). Occupational therapy and mental health. Churchill Livingstone Elsevier.

Kielhofner, G. (2009). *Conceptual foundations of occupational therapy practice* (4th ed.). Philadelphia: F. A. Davis Company.

Rivas-Garibay, A., Olvera-Romero, D., & Ferman-Cruz, G. (2011). Utilidad de la terapia ocupacional en pacientes psiquiátricos. *Revista Neurología, Neurocirugía y Psiquiatría, 44*(1), 13–17.

Smith, R., De Witt, P., Franzsen, D., Pillay, M., Wolfe, N., & Davies, C. (2014). Occupational performance factors perceived to influence the readmission of mental health care users diagnosed with schizophrenia. *South African Journal of Occupational Therapy, 44*(1), 51–55.

Sousa, S. (2006). Avaliação ocupacional na psiquiatria - tradução, adaptação cultural e validação da versão portuguesa da OSA (the ocupacional self assessment). Unpublished Master Thesis em Psiquiatria e Saúde Mental (Universidade do Porto). Porto, 2006.

Souza, F. N., Costa, A. P., & Moreira, A. (2011). Questionamento no Processo de Análise de Dados Qualitativos com apoio do software WebQDA. *EduSer - Revista de educação, 3*(1), 19–30.

Stemler, S. (2001). An overview of content analysis. *Practical Assessment, Research & Evaluation, 7*(17), 137–146.

Author Index

© Springer International Publishing Switzerland 2017
A.P. Costa et al. (eds.), *Computer Supported Qualitative Research*,
Studies in Systems, Decision and Control 71,
DOI 10.1007/978-3-319-43271-7

Printed in the United States
By Bookmasters